ADVANCES IN GENOME BIOLOGY

Volume 3B • 1995

GENETICS OF
HUMAN NEOPLASIA

ADVANCES IN GENOME BIOLOGY

Editor: *RAM S. VERMA*
Division of Genetics
The Long Island College Hospital-
SUNY Health Science Center
Brooklyn, New York

ADVANCES IN GENOME BIOLOGY

GENETICS OF

HUMAN NEOPLASIA

Editor: RAM S. VERMA

Division of Genetics

The Long Island College Hospital-

SUNY Health Science Center

Brooklyn, New York

VOLUME 3B • 1995

 JAI PRESS INC.

Greenwich, Connecticut   *London, England*

ISBN: 1-55938-835-8

Printed and bound by CPI Antony Rowe, Eastbourne

CONTENTS (Volume 3B)

CONTENTS (Volume 3A)

LIST OF CONTRIBUTORS

Peter A. Benn

Department of Pediatrics
University of Connecticut
Farmington, Connecticut

Johannes L. Bos

Laboratory for Physiological Chemistry
University of Utrecht
Utrecht, The Netherlands

Boudewijn M. Th. Burgering

Laboratory for Physiological
University of Utrecht
Utrecht, The Netherlands

Raju S. Chaganti

Cytogenetics Laboratory
Memorial Sloan-Kettering Cancer Center
New York, New York

Kathy Conway

Department of Epidemiology
Lineberger Comprehensive Cancer Center
University of North Carolina
Chapel Hill, North Carolina

Phillip M. Cox

Department of Histopathology
Royal Postgraduate Medical School
Hammersmith Hospital
London, England

Frank G. Haluska

Center for Cancer Research
Massachusetts Institute of Technology
Cambridge, Massachusetts

Daniel P. Heruth

Molecular Genetics Laboratory
The Children's Mercy Hospital
University of Missouri-Kansas City
Kansas City, Missouri

Bruce P. Himelstein

Division of Oncology
Children's Hospital of Philadelphia
Philadelphia, Pennsylvania

Catharina Larsson   Department of Clinical Genetics
                               Karolinska Hospital
                               Stockholm, Sweden

Ruth S. Muschel   Department of Pathology and Laboratory
                               Medicine
                               University of Pennsylvania School of
                               Medicine
                               Philadelphia, Pennsylvania

Eduardo Rodriguez   Cell Biology and Genetic Program
                               Memorial Sloan-Kettering Cancer Center
                               New York, New York

Paul G. Rothberg   Molecular Genetics Laboratory
                               The Children's Mercy Hospital
                               University of Missouri-Kansas City

Giandomenico Russo   Raggio-Italgene SpA
                               Rome, Italy

Thierry Soussi   Institut de Genetique Moleculaire
                               INSERM
                               Paris, France

Chandrika Sreekantaiah   Department of Pathology
                               New York Medical College
                               Valhalla, New York

Ram S. Verma   Division of Genetics
                               The Long Island College Hospital-SUNY
                               Health Science Center
                               Brooklyn, New York

Günther Weber   Department of Clinical Genetics
                               Karolinska Hospital
                               Stockholm, Sweden

Bernard E. Weissman   Department of Pathology
                               Lineberger Comprehensive Cancer Center
                               University of North Carolina
                               Chapel Hill, North Carolina

Jan Zedenius   Department of Clinical Genetics
                               Karolinska Hospital
                               Stockholm, Sweden

DEDICATION

To Donald F. Othmer and Mildred Topp Othmer with grateful appreciation for their commitment to the Long Island College Hospital, its research and cancer care activities, and for their financial support in establishing the Othmer Cancer Center.

PREFACE

The underlying idea that cancer is a genetic disease at the cellular level was postulated over 75 years ago when Boveri hypothesized that the malignant cell was one that had obtained an abnormal chromatin content. However, it has been only the last decade where enormous strides have been made toward understanding neoplastic development. Explosive growth in the discipline of cancer genetics is so rapid that any attempt to review this subject becomes rapidly outdated and continuous revisions are warranted. Conclusive evidence has been reached associating specific chromosomal abnormalities to various cancers. We have just begun to characterize the genes which are involved in these consistent chromosomal rearrangements resulting in the elucidation of the mechanisms of neoplastic transformation at a molecular level. The identification of over 50 oncogenes has led to a better understanding of the physiological process. Tumor suppressor genes, which were discovered through inheritance mechanisms, have further shed some light towards understanding the loss of heterozygosity during carcinogenesis. The message emerging with increasing clarity concerning specific pathways which regulate the fundamental process of cell division and uncontrolled growth.

The advances in molecular biology have led to a major insight in establishing precise diagnosis and treatment of many cancers resulting in prevention of death. The field is expanding so rapidly that a complete account of all aspects of genetics of cancer could not be accommodated within the scope of a single volume format. Nevertheless, I have chosen a few very specific topics which readers may find of great interest in hopes that their interest may be rejuvenated concerning the bewildering nature of this deadly disease. The contributors to Volume 3 have provided up-to-date accounts of their fields of expertise. Although the contributors have kept their chapters brief, they include an extensive bibliography for those who wish to understand a particular topic in depth.

For more than a century, cancer has been diagnosed on the enigmatic basis of morphological features. Establishing a diagnosis based on DNA, RNA, and proteins, which is done routinely now, was once inconceivable. Cloning a gene of hematopoietic origin is no longer a fantasy. The approach has shifted over the past 15 years from identification of chromosomal abnormalities toward zeroing in on cancer genes. The impact of new diagnostic technology on the management of cancer patients is enormous and I hope readers gain an overview on the progress concerning diagnosis and prevention.

I owe a special debt of gratitude to the distinguished authors for having rendered valuable contributions despite their many pressing tasks. The publisher and many staff members of JAI Press deserve much credit. My special gratitude to many secretaries for typing the manuscripts of various contributors.

Ram S. Verma
Editor

TRANSCRIPTION AND CANCER

Phillip M. Cox

I. INTRODUCTION

While every cell of a multicellular organism contains essentially the same complement of DNA in its genome, the various organs of its body perform vastly different

Advances in Genome Biology
Volume 3B, pages 233–278.
Copyright © 1995 by JAI Press Inc.
All rights of reproduction in any form reserved.
ISBN: 1-55938-835-8

functions. These different functions require the production of particular sets of proteins by the cells which necessitates expression of some genes and the silence of others. In addition, the levels of proteins expressed need to be altered in response to external stimuli, as cells progress through the cell cycle or as they develop from their embryonic precursors to their differentiated state.

Transcription is the process of copying small regions of the genomic DNA in the nucleus into mobile RNA species which, after processing, are translocated to the cytoplasm where the information they carry is translated into a polypeptide chain by ribosomes. Regulation of transcription is central to the control of the vast majority of cellular functions and, in particular, differentiation and the response of cells to external signals.

Cancer, or more generally, neoplasms, are clones of cells that have the ability to grow and divide independent of external controls and that, on the whole, are less differentiated than the cells of the organ in which they arose. Since transcriptional regulation is of such importance to both growth responses and differentiation, it is clear that deregulation of transcription must play an important part in the development of neoplasms.

Evidence is accumulating rapidly to support this notion, both from the study of the functions of viral oncogenes and their cellular counterparts and also from the genetic analysis of human tumors, most notably the leukemias.

II. EUKARYOTIC TRANSCRIPTIONAL REGULATION

In eukaryotic organisms, RNA transcription from the DNA template requires both the general transcriptional machinery, including RNA polymerase and basal transcription factors, and proteins (*trans*-acting factors) which specifically recognize short DNA sequences (*cis*-acting elements) usually in the noncoding region of the gene.[1,2] These sequences may be immediately upstream of the transcription start site where they constitute the promoter necessary for accurate and efficient initiation of transcription. Alternatively, they may lie hundreds or thousands of bases upstream or even within introns, exons, or downstream of the gene, forming enhancer elements capable of modulating promoter function.[3]

A. The General Transcription Machinery

RNA Polymerase II

In contrast to prokaryotes, which have a single RNA polymerase, eukaryotic cells contain three enzymes responsible for transcribing distinct groups of genes from the DNA template into RNA: (1) RNA polymerase I synthesizes the precursors of the large ribosomal (r)RNAs; (2) RNA polymerase II transcribes protein coding genes and some small nuclear (sn)RNAs; and (3) RNA polymerase III produces transfer (t)RNAs and the 5S rRNA.

RNA polymerase II (POLII) is a multisubunit protein complex of M_r ca. 500–600 kDa comprising two large polypeptides and 8–10 smaller components. Its exact composition *in vivo* has recently been elucidated and exhibits some variation depending upon the organism and promoter being studied.[4,5] The largest subunit, of M_r 220–240 kDa is highly conserved in structure throughout eukaryotes and shows considerable homology to the comparable subunits of RNA polymerases I and III and also to the β′ polypeptide of the prokaryotic enzyme.[6–10] The second largest subunit, M_r 140–150 kDa, also shows a high degree of evolutionary conservation in eukaryotes, and it shares many structural features with the β-subunit of its prokaryotic counterpart.[11,12]

The smaller components can be divided into three groups: first, those polypeptides used by all three RNA polymerases, and therefore presumably part of the fundamental core of the enzyme; second, those restricted to POLII but essential for its function; and third, those only found under some circumstances and at least partially dispensable. Although probably conserved among eukaryotes, these smaller polypeptides show only limited similarity to those in the prokaryotic enzyme.[4]

Through a combination of genetic and biochemical studies, both large subunits have been found to contact the DNA template and the nascent RNA chain.[13–16] In addition, the second subunit binds substrate nucleotides and appears to be involved in fashioning phosphodiester bonds, possibly with the assistance of one of the smaller polypeptides,[17] while a region on the largest subunit, to which the mycotoxin α-amanitin binds, is essential for RNA elongation.[18,19] The carboxy-terminal domain (CTD) of the largest subunit has multiple phosphorylation sites, which are unphosphorylated during the initiation of transcription when the exposed hydroxyl groups may interact with acidic regions of other transcription factors involved in regulating the rate of initiation at some promoters.[20,21] Phosphorylation, possibly by a general initiation factor (TFIIH), is essential for elongation to become established and may be involved in releasing the enzyme from the initiation complex.[22–26]

Basal Transcription Factors

In contrast to the bacterial RNA polymerase, which recognizes a specific DNA sequence a short distance upstream of the transcription start site, eukaryotic POLII depends upon a number of general transcription factors to locate the correct position for initiation. These were originally identified as activities in fractions of nuclear extracts separated by chromatography on phosphocellulose and necessary for reconstitution of a transcription system *in vitro*.[27] Originally three activities were defined: the transcription factors for RNA polymerase II (TFII) A, B, and D in the order of elution from the column,[27] with TFIIB subsequently being separated into

TFIIB and TFIIE.[28] The most extensively studied of these factors is TFIID, which interacts directly with an A–T-rich sequence in the promoter, the TATA box,[29,30] whose position (usually −25 to −30 relative to the transcription start site in mammalian genes) is critical in determining the exact point of initiation.[31] cDNA clones encoding yeast and mammalian TATA box-binding protein (TBP) have been isolated.[32–37] The protein encoded by the mammalian cDNA is considerably smaller than TFIID purified from nuclear extracts, which has been shown subsequently to be a multisubunit complex composed, in *Drosophila*, of six other polypeptides known as TBP-associated factors (TAF's).[38]

It is possible that several forms of TFIID coexist, comprising alternative combinations of subunits with slightly different binding specificities and distinct functions,[39] and allowing regulation of transcription via the TATA box. This may explain the earlier observations that some viral transactivators show specificity for the TATA boxes of particular genes[40] and that some TATA boxes are only functional in a particular promoter context.[41] It has been shown recently that TBP is also required for initiation on POLII promoters lacking a TATA-box[42] and, in addition, for transcription by RNA polymerases I and III,[43–46] where it appears to be associated with a different set of TAF's.

Binding of TFIID to the TATA-box is probably stabilized by interaction with a heterotrimeric complex, TFIIA,[30] and the resulting stable protein-DNA complex commits the template to transcription.[47,48] Upon this, a preinitiation complex is assembled by the addition of TFIIB, which probably interacts directly with TBP,[49] POLII, which is escorted by TFIIF, and finally TFIIE,[30] TFIIH, and TFIIJ.[26] The resulting massive structure is ready to initiate transcription and remains stable in the absence of nucleotides.[50,51] Initiation requires the hydrolysis of ATP or dATP, possibly by one of the TFIIE subunits,[52,53] and is associated with phosphorylation of the CTD of POLII by TFIIH [25,26] and a major conformational change in the preinitiation complex.[29,30] The released energy may be employed in unwinding the template around the initiation site to give access for the polymerase, creating a structure analogous to the prokaryotic "open complex", possibly with the assistance of TFIIE/F, which may possess helicase activity.[54]

Once the polymerase is released from the initiation complex and has transcribed a small number of nucleotides, it becomes part of an extremely stable elongation complex,[50,55] which probably includes at least some components of TFIIE as well as elongation factors.[56,57] This transcribes the whole of the gene, including the noncoding introns, and overshoots the end of the coding sequence, eventually terminating a considerable distance downstream. Meanwhile, TFIID and possibly some of the other parts of the preinitiation complex remain attached to the TATA-box facilitating the initiation of subsequent rounds of transcription.[29,55]

POLII and the general factors are necessary for the transcription of most, if not all, protein coding genes, with the rate and frequency of initiation being determined by additional *trans*-acting proteins which interact with specific *cis*-acting DNA elements in the promoter and enhancer regions. Thus, the combination of proteins controlling the transcription of an individual gene in a particular cell type depends (1) upon which *cis*-acting elements are present in the gene in question, and (2) upon which *trans*-acting factors are expressed and active within the nucleus.

B. Control of Eukaryotic Transcription

Structural Features of Transcription Factors

As a general rule, transcription factor proteins are organized as a series of either spatially discrete or occasionally overlapping peptide domains which mediate the various functions, such as binding to a specific DNA sequence, the interactions with other proteins responsible for dimerization, transcriptional activation, or repression and inactivation. As an increasing number of genes for transcription factors has been isolated, it has become clear that eukaryotes have evolved a fairly limited repertoire of structures capable of mediating these activities; thus transcription factors can be grouped together on the basis of shared structural features. Homology is particularly marked in the domains responsible for binding to DNA and for the formation of homo- or heteromeric complexes. In these regions, the amino acid sequence may show a high degree of conservation between proteins which are, otherwise, largely dissimilar. However, even single amino acid differences in these domains can change the DNA-binding specificity of a protein, either directly, or by altering the repertoire of other proteins with which it can form complexes.[58-65]

Two structures, the basic-helix-loop-helix (bHLH) and the leucine zipper (LZ), involved in dimer/multimer formation, have been extensively characterized and have some features in common. The bHLH motif was originally identified as a region required for dimerization and DNA-binding in three apparently functionally distinct groups of proteins with known or suspected transcriptional activity: the products of the *E2A* gene, E12 and E47, involved in, among other things, stimulation of B-lymphocyte-specific immunoglobulin gene expression; the muscle-specific factor, MyoD; *daughterless*, a *Drosophila* developmental gene product; and the Myc oncoproteins.[66] The same motif has since been identified in a wide variety of other proteins with diverse functions ranging from phosphatase gene regulation in yeast to peripheral nervous system development in *Drosophila*.[67,68] The bHLH domain comprises two predicted α-helices, separated by a peptide loop of variable length, and a 13-amino acid region containing a number of conserved basic residues, the basic domain, which lies immediately to the N-terminal side of the helices. The helices are believed to have hydrophobic amino acids positioned along one face (i.e., they are hydrophobic amphipathic helices) which mediate homo- or heterodimerization. Dimerization is required for DNA-binding by positioning the basic domains of the subunits such that they can bind to the cognate DNA sequence. Although the hydrophobic amino acids are essential for dimerization and are shared by all bHLH proteins, it is clear that other amino acids in the α-helices dictate specificity of dimer formation, since choice of a dimerization partner is highly specific.

The leucine zipper is an alternative dimerization motif, comprising a single proposed hydrophobic amphipathic α-helix with leucines on the hydrophobic face, through which related proteins interact.[69] As with the bHLH proteins, this helix may be positioned a specific short distance C-terminal to a basic domain which mediates DNA-binding, together forming the bZip domain.[70] This motif characterizes a growing family of transcription factors, several of which are capable of combining as stable and functionally active homo- and heterodimers.[69,71,72]

A separate group of proteins—the Myc proteins, and transcription factors AP-4 and USF—contains both a bHLH motif and a leucine zipper,[73] both of which are required for protein function. The LZ is positioned C-terminal to the bHLH, and probably cooperates with it in the formation of homo- or heteromeric complexes. Similarly, the LZ may be found in association with other DNA-binding motifs such as the POU-domain (see below).[74]

Two other frequently recurring motifs, the zinc finger (ZF) and the homeobox, have been shown to be directly responsible for sequence specific DNA-binding by separate families of transcription factors. The ZF motif was first described in the transcription factor TFIIIA, where it occurs nine times,[75,76] and up to 13 potential zinc chelation sites can be present in a single protein, as in a Y chromosome-derived protein originally believed to be the testis determining factor involved in sex determination.[77] It results from at least four appropriately spaced cysteine or cysteine and histidine residues forming a complex with a zinc ion. It was suggested that this would lead to the formation of a projecting "finger" by the amino acids between the two pairs of residues involved in the zinc complex which would interact with the DNA. X-ray crystallography has confirmed this structure in TFIIIA.[78] Although the requirement for zinc, shown by the loss of DNA-binding activity in the presence of ion chelating agents, has been demonstrated for several factors;[79] all zinc-dependent DNA-binding proteins are not structurally homologous. For example, while the "fingers" of Krüppel-like proteins interact with DNA, it is the amino acids between the two zinc chelation centers of the steroid receptors which make contact with the double helix.[59–61] In archetypal ZF proteins, such as Krüppel, the fingers are probably all functionally equivalent; in others they appear to subserve different functions. For example, Green and Chambon[80] showed that in members of the steroid hormone receptor family, which all have two chelation centers, one is needed for DNA-binding specificity, while the other stabilizes the protein–DNA complex, possibly by participating in protein–protein interactions with the other half of a dimeric receptor.[81,82]

The homeobox was originally identified as a highly conserved 180-base pair (bp) DNA element in the *homeotic selector* genes[83,84] involved in segmentation of *Drosophila*.[85,86] Subsequently, this structure has been found in a variety of transcriptionally active proteins from many phyla, many of which are involved in the processes of differentiation and development. The homeobox has been shown to mediate DNA binding by these proteins and to be responsible for sequence specificity.[59,87,88] Predictions of the structure of the homeobox, based on its amino acid sequence, suggested it would form three stretches of α-helix separated by short flexible spacers,[89,90] a motif very similar to the helix-turn-helix (HTH) first described in bacterial transcriptional repressors.[91] X-ray crystallography of the homeodomain of the *Drosophila* engrailed protein cocrystallized with DNA has provided final evidence of direct DNA–homeodomain interaction comparable with, although not identical to, that of the prokaryotic HTH. Helices 1 and 2 at the N-terminal end of the domain lie perpendicular to the DNA backbone and make few contacts, while helix 3 is inserted in the major groove of the DNA double helix, forming extensive contacts with the bases and sugar–phosphate backbone of the cognate sequence. In addition, amino acids N-terminal to helix 1 make further contacts in the minor groove.[92]

While for many proteins the homeodomain alone is sufficient for sequence-specific DNA binding, other related proteins require the presence of additional structures. One such group includes of the proteins Pit-1, Oct1 and 2, and Unc 86 (POU), the homeodomains of which are more closely related to one another than they are to an archetypal homeodomain, such as that of *Antennapedia(Antp)*. In particular, all share the amino acid sequence V-RVWFCN in helix 3 with V-RV–C-being POU-specific.[93] In contrast to the archetypal Antp and engrailed-like homeodomains, which alone are sufficient for sequence-specific DNA binding,[87,94] the POU family possesses a second highly conserved domain, the POU-specific box, which contributes significantly to specificity of DNA binding and which is required to generate high-affinity DNA contacts.[95,96] This POU-specific region comprises approximately 75 amino acids and is linked to the N-terminal end of the homeodomain by a short flexible spacer. Within the POU-specific box, two extremely well-conserved regions, subdomains A and B, are found, separated by a short segment of variable sequence and length. The four original POU proteins exhibit combined identity of 17/26 and 18/34 amino acids, respectively, over these two domains[93,97] and, in addition, three subserve similar functions being involved in the determination of cell fate.

The regions of transcription factors involved in transcriptional activation or repression are less well-conserved, although certain recurrent themes can be recognized. In particular, activation domains may contain a preponderance of amino acids with acidic side chains, or alternatively a single amino acid, such as proline or glutamine, may be especially abundant.[98] The relationship between the structure of these regions and their ability to modulate the rate of initiation of transcription by the RNA polymerase is not well understood.[99]

In general it is believed that activating domains function by promoting the assembly of the preinitiation complex. Consistent with this, the acidic activating domains have been shown, under different conditions, to form a stable interaction with either TFIID or TFIIB[100,101]; it has been suggested that the unphosphorylated "tailpiece" of POL II may be another target. In contrast, there is evidence that glutamine- and proline-rich activation domains interact with TFIID via intermediary coactivator proteins.[38,102,103]

The structural elements described above are not the only ones capable of forming dimers, of binding to DNA in a sequence specific manner, or of activating transcription. Other families are emerging whose conserved domains do not fit readily in any of the above models, but their structures are being solved and in time their mode of action will doubtless be determined.

Tissue-Specific Gene Expression

In a simple system, tissue-specific expression of a set of genes with related functions could be achieved by the production in the tissue of an active transcription factor capable of interacting with a *cis*-acting element shared by those genes. An example of this is found in the somatotroph cells of the adult anterior pituitary gland, which exclusively produce a factor, Pit-1/GHF-1, regulating expression of a number of pituitary-specific genes through a conserved DNA element.[104–108] Pit-1 appears to provide somatotroph specificity; however, expression of the various genes it regulates is subject to additional controls determining the level at which the genes are expressed.[109,110]

In general, the situation is not so simple since the same *cis*-acting element may be present in the regulatory region of genes which are not coordinately expressed.[111] Furthermore, several factors capable of recognizing that particular sequence may be present in the same cell.[112–115] To achieve tissue specificity, the sequence context in which the basic *cis*-acting element occurs is varied in different genes; the relative affinity of the factors for the core recognition site may be influenced by flanking DNA sequences, while their activity may be significantly modulated by protein–protein interactions with other sequence-specific DNA-binding proteins interacting with elements elsewhere in the same promoter/enhancer.[116,116a]

A major refinement of this relatively simple system has been achieved in eukaryotic organisms by the evolution of families of transcription factors which bind to their cognate sequence as homodimers or as heterodimers with other family members. By varying the combinations, different responses may be evoked from the same *cis*-acting element, and the range of elements recognized by a small number of factors can be extended, thus helping to facilitate the specific and flexible control of transcription required for normal cellular function.[66,71,117]

Inducible Gene Expression

While the mechanisms outlined above enable tissue-specific transcription of constitutively expressed genes, many genes are only activated in response to external stimuli. When a cell perceives such a stimulus, one of a number of signal transduction pathways may be activated, resulting in alteration in the level of transcription of responsive genes.[3] This may be achieved via the production of new transcription factors. For example, virus infection and exposure to IFNs α and β leads to induction of the synthesis of the transcription factors IRF-1 and -2, which bind to the promoters of IFN-inducible genes and to those of IFNs α and β themselves.[118-120] Alternatively, the DNA binding or transcriptional activity of factors already present within the cell may be modulated, either by covalent modification or as a result of interaction with other proteins or intracellular ligands.

Covalent modification may involve phosphorylation or glycosylation and, in addition, some factors are affected by changes in the redox state of their environment.

Phosphorylation, the best characterized of these mechanisms,[121] may regulate many functions of transcription factors including: nuclear localization;[122,123] DNA-binding, which may be either inhibited[124-126] or stimulated;[127-128] transactivation, also either inhibited[129] or stimulated;[130-132] and transrepression.[133] For some transcription factors the cellular kinases responsible have been determined, while for others they remain unknown. In addition, as exemplified by c-*jun*, phosphorylation of different sites on the same factor can regulate distinct functions. In the case of c-*jun*, glucose synthetase kinase 3 and casein kinase II-mediated phosphorylation of one set of sites, which are phosphorylated in quiescent cells,[126] inhibits DNA binding,[125] while phosphorylation of serine residue 63 or 73 stimulates transactivation in response to oncogenes.[131,132]

An alternative means of influencing transcription has been developed by families of factors which need to dimerize in order to bind DNA. By forming heterodimers with related proteins which share the dimerization motif, but which are unable to bind to DNA, potentially active factors can be sequestered and thus prevented from interacting with their cognate sequence.[134-137] The first such factor to be identified was IκB, an inhibitor of the factor NF-κB, but examples of this class of proteins have been found for the bZip (IP-1),[138] bHLH (Id),[135] and POU-domain (I-POU)[139] families.

Direct, ligand-mediated activation is a feature of the members of the steroid hormone receptor superfamily, all of which are probably transcription factors. In the absence of hormone, some of the receptors are sequestered in the cytoplasm by the heat-shock protein, hsp90. When the specific ligand enters the cell it occupies its binding site on the receptor protein causing a conformational change and release from sequestration, thereby allowing translocation to the nucleus, dimerization, and performance of its transcriptional function.[140-141]

A further level of regulation is achieved by the modulation of the binding or function of sequence-specific transcription factors by proteins which do not bind specifically to DNA in isolation but which may interact with or modify the transcriptional complex, inducing or repressing transcription. Such proteins include: the coactivators which perform a bridging function between some groups of transcription factors and the general transcription machinery; suppressor genes such as *rb1*, the retinoblastoma gene product; and a number of transactivating proteins encoded by DNA viruses such as E1a from adenovirus, VMW65 from Herpes simplex, and the SV40 large T antigen. In addition, as noted already, several groups of transcription factors are subject to phosphorylation by various cellular kinases.

By combining all of these mechanisms, expression of the vast array of genes in the eukaryotic genome can be regulated by a much smaller number of transcription factors.

Differentiation and Development

Besides being necessary for the normal function of mature cells, transcriptional regulation is central to the development of a normal body pattern and to the normal differentiation of embryonic cells into adult tissues. Genetic analysis in *Drosophila* of mutations leading to major structural abnormalities in the embryo or adult has identified genes whose products play a pivotal role in morphogenesis. Some 20 genes are involved in determining the dorsoventral axis of the early embryo,[142] while at least another 50 are required for segmentation of the embryonic body and the definition of structures arising from those segments. The genes involved in body segmentation encode proteins with features of transcription factors, which at specific times during the earliest stages of development, are expressed in a series of circumferential domains defining successively finer subdivisions of the embryonic body.[86] They show a hierarchy of regulatory interactions such that the first to be expressed, namely the *gap* genes, regulate the next class, the *pair rule* genes, and these in turn control the *segment polarity* genes. In addition, the members of each group modulate one another leading to the production of sharply defined segments. Unlike the *gap* and *pair rule* genes whose expression is transient, the *segment polarity* genes and the *homeotic selector* genes which they regulate, are persistently active, defining both the position and final fate of each cell.[86,143,144] Besides being controlled by genes higher up this regulatory cascade, the *homeotic selector* genes modulate one another and are subject to positive feedback. Thus, once activated their expression persists, providing the basis for programming of embryonic cells with their fate in the adult fly.

The cloning of several *homeotic selector* genes and subsequent structural analysis led to the identification of the homeobox DNA-binding domain[83] that is common to the products of virtually all genes of this class. Proteins with close structural similarity in man and mouse also show distinct domains of expression, and the homologues of the *homeotic selector* genes appear to define regional boundaries in the early development of the mammalian nervous system and mesoderm.[145-147]

Transcription factors are also of great importance in determining cellular fate by directing precursor cells along particular lines of differentiation. In the developing nervous system of the fruit fly and the nematode, *C. elegans*, a variety of such proteins have been shown to be required for differentiation of particular neurons,[148,149] while in mammals mutation of the gene encoding the POU protein Pit-1 disrupts the development of three of the five cell types in the anterior pituitary gland.[150]

Transcriptional Deregulation and Neoplasia

Transcriptional regulation is clearly fundamental to the normal response of cells to growth stimuli and for cellular differentiation. Disturbance of this regulation might therefore be expected to disrupt these processes and thus lead to the

Table 1A. Oncogenic Transcription Factors in Transforming Retroviruses

Family	Virus (Tumor)	Oncogene	Protooncogene	Mode of Activation
bZip				
	ASV17 (sarcoma)	v-*jun*	c-*jun* (AP-1)	deletion of repressor region
	FBJMuSV (OS)	v-*fos*	c-*fos* (AP-1)	deregulated expression
	AS42 (FS)	v-*maf*	c-Maf	NK
bHLH				
	MC29 (ML)	v-*myc*	c-*myc*	deregulated expression
ZF (Steroid)				
	AEV (EL)	v-*erbA*	T3 receptor	C' deletion generates constitutive repressor
Rel				
	ReVT (RE)	v-*rel*	c-*rel*	interferes with NF-κB family interactions
Myb				
	AMV (EL)	v-*myb*	c-*myb*	deletion of CKII-regulated repressor domain
	E26 (EL)			
Ets				
	E26 (EL)	v-*ets*	c-*ets1*	gag-myb-ets fusion protein
Ski				
	ALV	v-*ski*	c-*ski-1*	?

Note: OS-osteosarcoma; FS-fibrosarcoma; ML-myeloid leukemia; EL-erythroleukemia; RE-reticuloendotheliosis; NK-not known; T3-thyroid hormone.

Table 1B. Transforming Transcription
Factors Activated by Retroviral Integration

Family	Virus	Factor
ZF (Krüppel)		
	Ecotropic retrovirus Friend MuLV	Evi-1
(Other)		
	Moloney MuLV in Eµ-Myc mice	Bmi-1
Homeobox		
	Intracisternal A-particle (transposable element)	Hox 2.4
Ets		
	FeLV	Fli-1
	Spleen focus forming virus	Spi-1

uncoordinated growth and failure of differentiation seen in cancer cells. Deregulation might occur in a variety of ways; for example, transcription factors normally only expressed in response to external growth signals could be produced inappropriately or their usually short-lived mRNA or protein could be stabilized by mutation. Alternatively, mutation or deletion in the protein-coding sequence of such a transcription factor gene might result in a protein which was spontaneously active or which could not be suppressed. Since the amino acids forming the DNA-binding domain of a transcription factor determine the precise *cis*-acting element it recognizes and thus the range of genes it regulates, a single point mutation in this region could lead to activation of the wrong set of genes. At the same time, exchange of DNA-binding domains between unrelated factors as a result of genetic rearrangement could lead to activation by the chimeric factor of the correct target genes in response to the wrong stimuli. Abnormalities of proteins of the inhibitor class that sequester transcription factors in the cytoplasm could also have major effects, as of course could abnormalities elsewhere in the growth signaling mechanism which would be transmitted to the transcriptional machinery. Mutations could also affect transcription factors that normally mediate differentiation or that repress cell growth leading to their inactivation, or, alternatively, the genes for such factors might be disrupted or deleted.

In the ensuing sections, it will become clear that many of the mechanisms postulated above are responsible for the transforming activity of some retroviral oncogenes (Tables 1A and 1B) and that they may also be important in the development of human tumors (Table 2).

Table 2. Oncogenic Transcription Factors and Human Tumors

Family	Oncogene	Tumors	Mechanism
bZip	c-fos	Osteosarcoma	overexpression/
		Ca colon	gene amplification
	HLF	pre-B-ALL	chimeric factor
			[t(17;19)]
bHLH			
	c-myc	BL	activation [t(14;18)]
		(various)	gene amplification
	N-myc	Neuroblastoma	gene amplification
	L-myc	Ca bronchus	gene amplification
	E2	pre-B-ALL	chimeric factors
			[t(17;19) and t(1;19)]
	?max	?	? [t(14)]
	tal-1(TCL5)	T-ALL	activation [t(1;14)]/
			internal deletion
	lyl-1	T-ALL	activation [t(7;19)]
ZF			
(Krüppel)			
	wt	Wilms' tumor	inactivation (deletion
			or point mutation)
	gli-1	Glioblastoma	gene amplification
(Steroid)			
	RARA	Promyelocytic	chimeric factor
		leukemia	[t(15;17)]
(Other)			
	pml	Promyelocytic	chimeric factor
		leukemia	[t(15;17)]
	rhom1((ttg))	T-ALL	activation [t(11;14)]
	rhom2)		
Homeobox			
	pbx	pre-B-ALL	chimeric factor [t(1;19)]
	hox-11	T-ALL	activation [t(10;14)]
Rel			
	lyt-10	B-cell lymphoma	activation (truncated)
			[t(10;14)]
	bcl-3	B-CLL	activated
Other			
	p53	Many	Point mutation/deletion
	Rbl	Various	Inactivated by deletion

Note: Ca-carcinoma; ALL-acute lymphoblastic leukemia; BL-Burkitt's lymphoma; CLL-chronic lymphocytic leukemia.

III. TRANSCRIPTION FACTORS AS ONCOGENES

Since 1987, when, for the first time, an oncogene was shown to be a transcription factor, evidence has rapidly accrued that abnormalities of proteins with a direct role in transcriptional regulation can cause cellular transformation. This has come in part from the study of retroviral oncogenes and their relationship to cellular protooncogenes, and also from the cloning of genes involved in nonrandom chromosome rearrangements found in a variety of human tumors.

Oncogenes, originally identified as the transforming genes of highly oncogenic retroviruses,[151] were later recognized as modified forms of normal cellular genes — so-called protooncogenes, the products of which are involved in the control of cell growth and proliferation. Inappropriate expression of these genes results in transformation of cells in culture.

Although many protooncogene products are growth factors, surface membrane receptors and cytoplasmic or membrane-bound proteins (for review see, Ref. 152), a number were localized to the nucleus and it was proposed that they provided the final stage in the growth signaling pathway.[153] Comparison of the functions of the normal protooncogene product with its transforming retroviral counterpart has provided some valuable insights into the ways in which cellular proteins can be rendered oncogenic. However, to date, the large majority of retrovirus-induced neoplasms have been found in animals and birds and few abnormalities of these genes have been found in human tumors.

In many human tumors certain chromosomal rearrangements occur at a much higher frequency than would be expected by chance, suggesting that the genetic lesions caused by such nonrandom events are important for development of the neoplasm. Cloning of the genes adjacent to the breakpoints of such rearrangements has demonstrated in a number of cases the involvement of a transcription factor or factors.

Currently, some 30 proteins with known or presumed transcriptional activity, including members of all the major structural families, have been implicated in cellular transformation.

A. bZip-Family Oncoproteins

In 1987, Maki et al.[154] were first to demonstrate a direct relationship between oncogenes and transcription factors. They demonstrated that the product of v-*jun*, the transforming gene of the chicken sarcoma virus ASV17, contained a region of considerable amino acid sequence similarity with the yeast transcription factor GCN4 across from its DNA-binding domain.[155] When the DNA-binding domain of GCN4 was replaced with the homologous region of v-Jun, the chimeric protein continued to function as a transcription factor,[156] implying that the v-Jun protein would also bind DNA.

The DNA sequence motif recognized by GCN4 — (**ATGA(C/G)TCAT**)[157] (A = adenosine, C = cytidine, G = guanosine, T = thymidine) — is very similar to the cognate sequence, **TGACT(C/A)A**, of AP-1, a transcription factor (or family of factors) involved in mediating transcriptional activation in response to stimulation of protein kinase C.[158,159] Monoclonal antibodies raised against different regions of v-Jun were shown to precipitate AP-1, but not other transcription factors, from nuclear extracts[160] and subsequently the peptide sequence of a 47-kDa protein in purified AP-1 preparations was shown to be identical to c-Jun.[161]

When compared with c-Jun, v-Jun shows two amino acid changes in the DNA-binding domain, an internal deletion of 27 amino acids towards its N-terminus, and its mRNA lacks a large 3′ untranslated region. Normal c-Jun shows only weak transforming ability in a focus forming assay, but creation of the N-terminal deletion present in v-Jun increases the number of foci formed 10-fold. Removal of the 3′ untranslated sequences also has some effect, probably as a result of mRNA stabilization, while the point mutations appear to have no role in transformation.[162]

Although v-Jun and c-Jun show similar affinity for AP-1 DNA-binding sites, v-Jun is a considerably stronger transcriptional activator in HeLa cells. This is a direct result of the internal deletion since when this region, designated δ, is removed from c-Jun its activating ability is also enhanced, suggesting that δ acts as a repressor domain. In other cell types, however, the repressor function of δ is not evident. This appears to be due to the absence of an additional cell-type specific factor (the δ repressor) which mediates the repression by interacting with δ.[163] Since loss of cell-type specific repressor function and transforming ability of v-Jun reside in the same region, it is likely only cells carrying the δ repressor will prove transformable by v-Jun.

The transforming gene of the Finkel-Biskis-Jinkins murine sarcoma virus (FBJ-MuSV) is v-*fos*.[164] Its cellular counterpart, c-*fos*, like c-*jun*, is an immediate early gene, whose product is a nuclear phosphoprotein expressed transiently in many cell types in response to mitogenic and other stimuli.[165] It was widely believed to be a transcription factor; however, sequence-specific DNA-binding could not be demonstrated. However, immune precipitation experiments showed that c-Fos was present in complexes binding to the AP-1 sequence.[166] Furthermore, a number of other proteins were coprecipitated as a result of interaction with Fos, the most abundant of which was c-Jun.[167,168]

Like Jun, Fos has a region of homology with the yeast factor GCN4, and analysis of the likely protein structure of this conserved domain led to the recognition that each contained a leucine zipper (LZ), first identified in C-EBP[70], which facilitates the protein–protein interaction between Fos and Jun necessary for DNA binding.[69,70] Unlike Jun, which can form homodimers with, albeit relatively weak, DNA-binding and transactivating ability, Fos is unable to homodimerize. However, Fos–Jun heterodimers are stable, bind to AP-1 sites with high affinity, and are potent activators of transcription.[71]

In its C-terminal 49 amino acids, v-Fos differs from c-Fos as a result of a deletion which alters the reading frame of the viral gene.[164] However, this has no significant effect upon the ability to form specific DNA-binding heterodimers with Jun, or to activate transcription from TPA-responsive elements;[168,169] both will transform fibroblasts.[170] Indeed, mutations in the LZ of v-Fos which prevent it from heterodimerizing also destroy its transforming ability.[171] Therefore, it appears that it is loss of normal regulation of Fos protein expression and not the altered C-terminus, which is responsible for the transforming ability of v-Fos. Overexpression of Fos mRNA and protein has been detected in a number of tumors, although it is not clear whether this is a primary event or the result of growth stimulation, and amplification of the c-*fos* gene is also occasionally detected.[172,173]

The bZip domain present in Jun and Fos has been identified in a large family of proteins, including others which bind to AP-1 sites and also those recognizing ATF/CREB sites which mediate cAMP responses,[174-178] and many have been shown to be capable of selectively forming heterodimers.[69,71,72] It is also clear that some AP-1 and ATF/CREB-binding factors can heterodimerize, thus there is a considerable repertoire of heterodimers able to bind to the same, or closely related, DNA sequences under different conditions.

A further level of complexity in the function of AP-1 and possibly other bZip proteins was discovered when it was shown that there is a close relationship with members of the steroid receptor superfamily,[179] which are also transcription factors. A small number of genes have *cis*-acting elements capable of binding both steroid receptor and AP-1, either separately or simultaneously, with the transcriptional response being dependent upon which factor(s) are present.[180,181] In addition, AP-1 and possibly ATF can repress promoters containing glucocorticoid or retinoic acid receptor-binding sites without an AP-1 site, and vice versa.[182-184] This effect is the result of interaction between the zinc fingers of the steroid receptor and the zipper of the bZip protein, most probably causing mutual interference with DNA binding. Since both v-Fos and v-Jun have an intact LZ it is possible that some of their effects could be produced in this way.

Despite the large size of the bZip family, only one other member has been found to be a viral oncogene, v-*maf*, which causes naturally occurring fibrosarcomas in fowl, has both a leucine zipper motif,[185] and an adjacent domain of basic amino acids which fits the consensus sequence of the bZip proteins.[70] In addition, v-*maf* has stretches of uninterrupted glycine and histidine residues, possibly representing regulatory domains.

Another gene for a bZip protein, *HLF*, is disrupted by a t(17:19) translocation described in two patients with acute lymphoblastic leukemia (ALL) showing a primitive B-cell phenotype.[186] The part of the gene encoding its bZip domain becomes fused to the gene *E2A* on chromosome 19. *E2A* codes for two HLH proteins, E12 and E47, which recognize a common enhancer element and are active in B cells. A chimeric protein is produced, with the HLF bZip domain fused to the E12/E47 activation domain and expressed under the *E2A* promoter.

The normal HLF protein is most closely related to TEF, a bZip protein found in embryonic thyrotroph cells of the anterior pituitary gland, and DBP, which regulates albumin gene expression in the adult liver. HLF is expressed in adult liver and kidney, but not in lymphoid cells; thus the chimeric protein may interfere with normal bZip or bHLH protein-mediated transcriptional regulation of differentiation in B cells.

B. Basic Helix-Loop-Helix Oncogenes

The oncogene v-*myc*, carried by the avian retrovirus MC29, was one of the earliest viral transforming genes to be identified. The protein products of its cellular counterpart, c-*myc*, and the related L- and N-*myc* genes, have features of immediate early genes, being expressed in the nucleus[187,188] and showing a rapid response to growth stimuli.[189] Inappropriate expression of different members of the *myc* family is found in a number of human tumors, either from an apparently normal *myc* gene, or as a result of chromosomal translocation, such as the t(14;18) found in Burkitt's lymphoma, or gene amplification, which is commonly present in neuroblastoma, lung, breast, and cervical carcinoma.[190–193] In addition, like *fos*, *myc* can complement *ras* oncogenes in transforming cells.[194]

Although nonspecific binding to DNA was demonstrated,[187] little more was known about the function of these proteins until a region of amino acid sequence homology was recognized between the *myc* gene products and a number of known and putative transcription factors involved in differentiation and tissue-specific gene expression. In these transcription factors, this domain, the bHLH motif, mediates the formation of homo- and heterodimers between related proteins and is essential for sequence-specific DNA binding.[66,195,196] Thus by inference *myc* was also likely to have a role in transcription. To the C-terminal side of the bHLH domain Myc, in common with the transcription factors AP-4 and USF, also has a leucine zipper domain, similar to that present in Fos and Jun. This has also been shown to be necessary for the formation of Myc multimers, but which also may enable interaction with other proteins.[197]

Given the homology with other members of the bHLH family, it was clear that Myc should bind DNA. Two complementary approaches resulted in identification of the Myc recognition sequence. First, by selection and PCR amplification of DNA interacting with Myc protein from a pool of random sequences a consensus target site, CACGTG, was identified.[198] Second, on the basis of homology across the basic regions of Myc and the yeast transcription factor PHO4, which binds the sequence CACGTG, it was demonstrated that a chimeric protein comprising the Myc basic domain substituted for that of PHO4 would bind the CACGTG motif *in vitro*.[62] These experiments, together with those from a number of other laboratories showed that Myc was potentially a sequence-specific DNA-binding protein. However, since DNA binding by Myc was extremely inefficient, requiring large amounts of protein to produce binding, it was suggested that Myc might function as a heterodimer.

A dimerization partner for c-Myc was isolated from a λgt11 expression library, by screening with the dimerization domain of c-Myc.[199] This protein, Max, exists in two forms differing by the presence of an additional 9 amino acids N-terminal to the basic domain in the larger form. It can homodimerize and heterodimerizes with c- L- and N-Myc, both *in vitro* and *in vivo*, and both homo- and heterodimers recognize the CACGTG consensus.[200,201] Max cannot heterodimerize with other bHLH/LZ proteins such as AP-4[199,202], but recently a second partner for Max, Mad, has been identified (in press). The structure of Max is simple, comprising 160 amino acids, half of which constitute the bHLH and LZ regions responsible for dimerization, flanked by a short N-terminal domain and a C-terminal region which includes a small acidic domain. In contrast to c-Myc, Max expression is seen in quiescent as well as dividing cells and thus it may act to repress transcription from Myc-responsive genes in the absence of c-Myc protein. The ability of Myc:Max heterodimers to transactivate gene expression is dependent upon the N-terminus of Myc, while Max homodimers repress transcription from the same genes[203] and DNA-binding of Max homodimers, but not Myc:Max heterodimers, is prevented by phosphorylation of Max by casein kinase II.[204] Transcriptional activation is clearly important for the transforming activity of Myc proteins since the long N-terminal domain responsible for transactivation[205] is required for cell transformation,[206] and the relative potency of the activation domains of c- and L-Myc appears to correlate with their transforming ability in rat embryo cells.[207] It is also possible that Max may be directly implicated in human neoplasia, since the *max* gene has been localized to chromosome 14 bands q22-24, a region subject to deletion in B-cell chronic lymphocytic leukemia and close to the breakpoint of the 12;14 translocation found in uterine leiomyomas.[208,209]

Three additional bHLH proteins have been implicated in different forms of acute leukemia. One of these, the product of the *E2* gene, participates in separate translocations in pre-B type acute lymphoblastic leukemia with a bZip protein (see above) and with a homeobox gene (see below). The others, Tal-1(SCL/TCL5) and Lyl-1, are closely related and both are expressed in hematopoietic cells. The *tal-1* gene was identified by cloning of the breakpoint on chromosome 1 caused by the translocation, t(1:14) (p33;q11.2) in a primitive acute leukemia capable of both myeloid and lymphoid differentiation. The normal *tal-1* gene is expressed by hematopoietic stem cells and myeloid and T-lymphocyte precursors,[210] and may also be important in erythropoiesis.[211] Its encoded protein can interact with other bHLH proteins forming heterodimers which specifically recognize the E-box DNA motif, CANNTG, found in a number of enhancers. The translocation results in an aberrant mRNA, including at its 3′ end part of the T-cell receptor (TCR) δ-chain gene. A small group of T-cell acute lymphoblastic leukemias (T-ALL) carry the translocation t(1:14) (p32;q11) in which *tal-1* is also involved,[212] and up to 25% of T-ALLs carry a small internal deletion in this gene.[213] Furthermore a rearrangement of *tal-1* has been detected in a human melanoma cell line.[214] The T-ALL translocation results in truncation of the *tal-1* gene and fusion to the 5′ part of the *tcr*δ gene, and thus the function of the protein may be abnormal and it may be expressed under the control of an inappropriate promoter.

The second protein, designated Lyl-1, is expressed in a truncated form in another subset of T-ALLs, as the result of a t(7:19) translocation which juxtaposes its gene to that of the TCR β-chain.[215] The normal Lyl-1 protein is expressed in lymphoid cells; however, its function is not known.

Further study of these proteins should give valuable insight into the role of transcription in normal hematopoietic development and in lymphoid neoplasia.

C. Zinc-Binding Transcription Factors as Cancer Genes

The archetypal zinc-dependent DNA-binding domain is the zinc finger, present in TFIIIA and the *Drosophila* protein Krüppel. However, other families of transcription factors use the chelation of zinc ions to form different structures.[82] For example, the steroid receptors bind DNA through a predicted α-helix adjacent to the zinc chelation site, while in several fungal transcription factors a third structure has been defined. Other less well-characterized groups also exist for which the nature of the DNA-binding structure has not been determined precisely. Members of the various groups have been shown to be either dominant positive, dominant negative, or recessive oncogenes.

The Wilms' Tumor Gene and Other Krüppel-Like Proteins

Wilms' tumor (WT) or nephroblastoma is an embryonal tumor of the kidney, which occurs sporadically in young children, but usually in a few cases it is hereditary or is associated with other congenital abnormalities including aniridia, defective genito-urinary development, and mental retardation (the WAGR syndrome). Gross cytogenetic deletions and rearrangements affecting chromosome 11(band p13) are a common finding in the WAGR syndrome and in occasional sporadic Wilms' tumors, a finding which raised the possibility that this region contained a recessive oncogene (tumor suppressor gene). By cloning and sequencing the breakpoints of these rearrangements a gene was defined which is consistently disrupted by these changes. This proposed *wt* gene encodes a protein (WT1) with four "zinc fingers" of Krüppel-type near to its C-terminus, suggesting that it is a transcription factor.[216,217] Its N-terminus contains proline- and glutamine-rich domains, reminiscent of the activation domains of some other transcription factors.[218,219] On the basis of amino acid conservation, WT1 is most closely related to EGR-1,[220] an early growth response gene product which recognizes the DNA sequence element 5′-CGCCCCCGC-3′ and acts as a transcriptional activator.[221,222] WT1 also recognizes this element but functions instead as a repressor.[223] The *wt* gene is strongly expressed in the embryonic kidney where its product appears to be required for the switch from mesodermal to epithelial differentiation that occurs in the developing urogenital system.[224] In addition, it is widely expressed in both male and female genital tract in the embryo and adult.[225] Besides the gross cytogenetic abnormalities occasionally seen, a proportion of sporadic Wilms' tumors show loss-of-heterozygosity at this locus or carry small internal deletions or point mutations in the *wt* gene, providing support for its role in tumorigenesis. Furthermore, nephroblastoma cells transfected with a normal chromosome 11, are unable to form tumors in mice.[226] The mechanism by which homologous deletion or rearrangement of the *wt* gene allows uncontrolled proliferation of nephroblasts is unknown, but one might speculate that the normal WT1 protein stimulates their differentiation into nephrons, possibly by repressing genes which stimulate growth and which remain active when *wt* is disrupted. One candidate target for WT1-mediated repression is the insulin-like growth factor II (IGFII) gene, whose product is consistently overexpressed in Wilms' tumors and which is normally repressed by WT1.[227]

Other Krüppel-like zinc finger proteins have been shown to be dominant oncogenes. The gene *gli1*, on chromosome 12, is amplified in a proportion of human malignant gliomas.[228] Evidence suggests that the GLI1 protein may be involved in normal development since it is most closely related to the Krüppel zinc finger proteins[229–231] involved in *Drosophila* segmentation and is produced by embryonal carcinoma cells but not by adult tissues. In addition, disruption of the very similar *gli3* gene is responsible for Greig's syndrome, a disorder of craniofacial and limb development.[232]

Another ZF gene, *evi-1*, is activated by integration of ecotropic retrovirus or by the Friend murine leukemia virus in a number of interleukin-3 (IL-3)-dependent myeloid leukemia cell lines.[233–235] It encodes a 120-kDa protein containing 10 Krüppel-like ZF's and an acidic putative activation domain. Bacterially expressed, *evi-1* specifically recognizes the consensus sequence TGACAAGATAA which is found in the regulatory region of a number of genes.[236] High-level expression is seen in a small number of specific regions in the embryo and exhibits temporal regulation, suggesting that it plays an important developmental role, while widespread low-level expression is found in adult tissues.[237]

The Steroid Receptor Family

The best characterized of the oncogenes belonging to the steroid receptor class of proteins is v-*ErbA*, one of the transforming genes of the avian erythroblastosis virus which encodes a truncated thyroid hormone receptor,[238,239,242] a relative of the steroid hormone receptors.[240,241] As a result of a small C-terminal deletion, the viral ErbA protein shows no affinity for the thyroid hormone T3 in mammalian cells. It does, however, bind correctly to thyroid hormone-responsive DNA elements, functioning as a constitutive repressor of thyroid hormone responsive genes and competitive antagonist of the normal thyroid hormone receptor/ligand complex,[242] a capacity which correlates with transforming ability.[243] This implies that v-ErbA may transform erythroblasts by blocking thyroid hormone-mediated differentiation, placing it in the dominant negative class of oncogenes. This may not be the complete story since when v-ErbA is expressed in yeast cells it is able to respond to thyroid hormone, suggesting that other factors may influence it in mammalian cells.

In acute promyelocytic leukemia (APL; FABM3), tumor cells carry the reciprocal translocation t(15:17)(q22;q11.2-q12) in approaching 100% of cases.[244] The chromosome 17 breakpoint has been characterized in a number of cases and shown to fall in the first intron of the retinoic acid receptor α-gene (*RARA*),[245] whose product is a member of the steroid hormone receptor family. The translocation separates the first exon, which encodes the transcription activation domain, from the DNA-binding and ligand-binding domains which become part of the derivative chromosome 15. The gene to which they are attached (*pml/my1*) is a member of a third group of proteins with a cysteine-rich, proposed zinc-binding domain.[246] This structure appears distinct from the Krüppel-like and steroid receptor zinc finger motifs and probably mediates DNA binding since the family comprises known yeast and mammalian transcription factors, *Drosophila* developmental genes and a recombination-activation protein involved in immunoglobulin gene rearrangement.[247–249] *pml* is a myeloid specific gene and is presumably freed of its normal regulation by the translocation, causing it to become a dominant oncogene; however, the fusion protein is responsive to retinoic acid and high dose RA therapy will cause remission in patients with APL, possibly by inducing myelocytic differentiation.

Other oncogenic members of the same group of cysteine-rich, putative zinc-binding transcription factors such as *PML* have been described, including *bmi-1* and *mel-18*. The gene *bmi-1* is a common site of integration for the Moloney murine leukemia virus in a transgenic mouse line carrying an activated c-*myc* gene, expressed under the control of an immunoglobulin heavy chain enhancer. The cooperative effect of *myc* and *bmi-1* leads to the development of B-cell lymphomas with a much shorter latency than those formed with c-*myc* alone.[250,251] The closely related *mel-18* is a murine neural crest-specific gene expressed at high level in melanoma cells and other tumors, and thus is suggested to play a role in cell proliferation.[252,253]

Two T-ALL-associated translocations involving chromosome 11, bands p15 and p13, activate expression of the genes *rhom-1* and *rhom-2*, respectively, both of which encode proteins with cysteine-rich putative zinc-binding LIM domains related to the ZF.[254–256] Both genes are expressed in the embryonic mouse nervous system and, in addition, *rhom-2* is present in the developing lung, kidney, and spleen.[255,257] Presumably aberrant expression of either of these proteins in T cells as a result of the translocations interferes with normal differentiation or stimulates T cell proliferation.

It is clear that the proteins described above are not likely to be the only developmentally regulated transcription factors with a role in transformation, since any such protein which can promote replication of embryonic cells or maintain such cells in an undifferentiated state would have equally serious consequences if expressed inappropriately in adult tissues.

D. Homeobox Genes

The homeobox motif is widely distributed in genes with roles in development and cellular differentiation;[86,105,258,259] however, few examples of such genes being oncogenic have been identified.

In childhood pre-B-acute lymphoblastic leukemia (pre-B-ALL), the tumor cells of 30% of patients carry the translocation t(1:19)(q23:p13.3).[260] The translocation results in a chimeric gene from which a potentially functional transcription factor is synthesized,[261,262] with the N-terminal part of the novel protein deriving from the *E2A* gene on chromosome 19 which encodes the Myc-related proteins E12 and E47. These are bHLH proteins capable of heterodimerizing with a variety of other family members forming enhancer binding factors active on a variety of genes including the kappa immunoglobulin promoter in B cells.[196,263] The part of the protein retained in the chimeric factor is the potent transcriptional activator domain. The C-terminus is encoded by *pbx*, a gene on chromosome 1, normally inactive in pre-B lymphocytes, which contributes its homeodomain. The result of the pre-B-ALL translocation is a protein which may be capable of strongly activating a gene or group of genes involved in the control of normal differentiation and not normally expressed at this stage of lymphocyte ontogeny.

The homeobox gene *hox-2.4*, is constitutively expressed in the mouse myeloid leukemia cell line WEHI-3B as the result a rearrangement caused by insertion of a transposable DNA element of the intracisternal A-particle type.[264] The oncogenic potential of this gene may extend beyond haematopoietic cells since when the activated *hox-2.4* is introduced into NIH3T3 fibroblasts the resultant cells form fibrosarcomas in nude mice.[265]

In approximately 5% of T-ALLs the translocation t(10;14)(q24;q11) is present, which juxtaposes the tcr-δ chain gene with another homeobox gene, *hox11* (*tcl3*), expressed in normal liver, but not in the thymus or in T lymphocytes.[266–268] It is presumed that deregulated expression of *hox11* interferes with the normal process of T-cell development and thus causes transformation.

E. *rel* and its Relations

ReVT is an avian retrovirus which causes rapidly progressive lymphoid tumors (reticulendotheliosis) in birds. It carries the v-*rel* oncogene[269] whose product is a protein found in the nucleus of some cells and in the cytoplasm of others.[270] The protooncogene c-*rel* is one of a number of genes related to NFκB, a multifunctional transcription factor, and *dorsal*, a *Drosophila* gene involved in dorsal/ventral axis determination.[271,272] NFκB is a heteromeric complex of two polypeptides of the same family, p50 and p65. p50 homodimers can bind to κB elements (as the factor KBF1) while p65 does not bind DNA on it own. In cells where it is not constitutively active, NFκB is sequestered in the cytoplasm through the interaction of p65 with IκB, another relative,[133] which in turn may be attached to the cytoskeleton via a series of "ankyrin-like" repeats. p50 is produced from a 105-kDa non-DNA binding precursor (p105) also having "ankyrin-like" repeats. Its activity may be regulated partly through the rate of cleavage of this cytoplasmic molecule to p50, which can then localize to the nucleus, associate with p65, and bind to DNA.[271,272] Nevertheless, sequestration of p50 in the cytoplasm by IκB is probably the more important regulatory event.

The v-*rel* oncogene differs from c-*rel* by 14 base substitutions in the Rel-homology domain responsible for DNA-binding and dimerization. It also lacks approximately 100 C-terminal amino acids. The deletion interferes with cytoplasmic retention, but replacement of the missing amino acids does not affect transforming ability.[270,273] The c-*rel* and v-*rel* oncogenes are both able to interact with other NFκB-like factors; however, v-*rel* is a much weaker activator of transcription. This suggests that it may interfere with the normal equilibrium between active and inactive complexes, possibly by acting as a dominant negative transcriptional repressor, or alternatively, by sequestering other important regulatory proteins. For example, v-*rel* can abolish the powerful stimulatory effect of phorbol esters on a reporter plasmid controlled by a κB site,[274] and it is proposed that v-*rel* forms nonfunctional heterodimers with other family members. From this point of view, v-*rel* and v-ErbA appear to function in a similar fashion. Interference with Rel family interactions appears to be the mode of action of a naturally occurring differential splice product of the p65 NFκB subunit, p65Δ, which lacks amino acids necessary for heterodimerization with p50 and DNA binding. This protein is highly expressed in proliferating hematopoietic cells and will transform RAT-1 fibroblasts,[275] presumably employing a similar mechanism to v-*rel*.

Studies of human lymphoid neoplasms have revealed abnormalities of two other members of the extended Rel family. The first is *lyt-10*, a gene encoding a protein containing a Rel-homology domain and "ankyrin-like" repeats, which is juxtaposed to an immunoglobulin locus by a t(10:14) translocation in a B-cell lymphoma.[276] In the normal Rel-related proteins the "ankyrin-like" repeats perform a regulatory function, preventing DNA-binding either by cytoplasmic sequestration or by physically obstructing the dimerization domain until they are removed by proteolytic cleavage. The translocation results in a Lyt-10 protein lacking this region and thus possessing constitutive transactivating activity.

The second gene is *bcl-3*, which is overexpressed in B-cell chronic lymphocytic leukemia. The product of this gene has homology with IκB, containing six "ankyrin-like" repeats, and can inhibit the DNA-binding activity of both NFκB and KBF1 *in vitro*,[277] although its function *in vivo* is as yet undetermined.

This extended protein family is likely to prove of continuing interest for some time, as its complex interactions are unraveled and the functions of the various family members are determined.

F. *myb* and *ets*

v-myb

Two chicken retroviruses, avian myeloblastosis virus (AMV) and E26, carry the oncogene v-*myb*, a truncated version of the cellular protooncogene c-*myb*,[278,279] which encodes a nuclear protein of M_r 75–80 kDa[279–281] expressed in immature, but not in differentiated hematopoietic cells.[282] Both v-Myb proteins lack the N-terminus and most of the first of three N-terminal repeats responsible for DNA-binding, plus a large part of a C-terminal negative regulatory domain.[283,284] In addition, in E26, v-Myb is expressed as a fusion protein with v-Ets.

When chicken bone marrow cells are transformed with v-Myb their differentiation is blocked and they have the phenotype of myeloid precursors,[285] while overexpression of c-Myb prevents the induction of differentiation of cultured erythroleukemia cells.[286] Furthermore, transformation of myeloid cells with v-Myb leads to dedifferentiation, thus *myb* expression appears incompatible with a differentiated state.

v- and c-Myb bind specifically and with similar affinity to the sequence PyAACG/TG (Py = pyrimidine) and activate transcription from reporter genes linked *in cis* to this motif.[283,284,287] The structure of the DNA-binding domain has been determined, and comprises a tryptophan-based HTH-like domain.[288,289] *Myb* activates *mim1*, encoding a promyelocyte-specific secretory protein,[290] and it may also instigate an autocrine growth loop by increasing expression of insulin-like growth factor-1 and its receptor.[291] Other targets of Myb activation are not known.

Phosphorylation at a number of serine residues in the N-terminus of c-Myb by casein kinase II has a negative effect on transcriptional activation, by preventing DNA-binding. The region of c-*myb* containing these residues is deleted from v-*myb* and activated c-*myb* genes,[124] thus transformation may be the result of expression of a Myb protein which inappropriately activates transcription because it cannot be prevented from binding to DNA. Alternatively, the disruption of the C-terminal negative transcriptional domain, which results from some retroviral integrations, can also cause transformation, while retroviral vectors carrying C-terminally truncated Myb will immortalize myelomonocytic precursors, albeit at low frequency. Most probably optimal transforming ability requires both truncations. It is not clear whether loss of part of the first repeat of DNA-binding domain is important for transformation; however, a subtle effect on DNA-binding specificity cannot be excluded. Further elucidation of the mechanism by which C-Myb is rendered oncogenic and the genes responsible for the dedifferentiated state of transformed cells should be most illuminating.

The Rapidly-Growing ets Family

The second oncogene present in the genome of avian leukosis virus, E26, which also bears v-*myb*, is *v-ets* and both genes are necessary for it to induce erythroblastosis.[292–294] The product of the chicken protooncogene, c-*ets-1*, from which it is derived, is a member of a family of nuclear phosphoproteins of short half-life.[295–297]

Related proteins all sharing a highly conserved region of amino acids with overall positive charge at the C-terminus (the ETS-domain)[298] are also present in insects and other vertebrates.[299-301] The c-*ets-1* gene binds specifically to a *cis*-acting element centered on the motif GGAA, found in the long terminal repeat (LTR) of the Moloney murine sarcoma virus[302] and HTLV1,[303] in the closely related PEA3-motif, in the polyoma virus enhancer, in the promoters of extracellular matrix-degrading enzymes,[304] and in the T-cell receptor-α enhancer.[305] Ets activates transcription from these sequences and, in the polyoma enhancer, functions cooperatively with members of the Fos–Jun family which bind to an adjacent AP-1 site.[306] Juxtaposed AP-1 and PEA3 sites are a feature of a number of oncogene-responsive promoters[307] and may represent an important target for extranuclear oncogenes such as *ras*.

Phosphorylation of c-*ets-1* and -2 by a cellular kinase, possibly the protooncogene product *raf-1*, inhibits DNA-binding.[297,308] The expression of c-*ets-1* is high in quiescent T cells and repressed on T-cell activation,[309] while it is induced in proliferating endothelial cells.[310]

The v-Ets oncoprotein has limited independent transforming ability, and coexpression with Myb has an additive effect; however, the E26 Gag-Myb-Ets fusion protein, has much greater potency, suggesting that it possesses some additional properties over those of the individual oncoproteins which are its constituent parts.[311]

The genes for two other ETS-domain proteins, *fli-1* and *spi-1* are activated by integration of two different retroviruses,[312,313] while other family members interact with the serum-response factor (SRF) on the promoter of serum-responsive immediate early genes such as c-*fos*.[314,315]

The Ets family members regulate a number of viral and cellular genes and may have important roles in both differentiation and responses to external growth signals, while the expression of c-*ets-1* in proliferating endothelial cells and its regulation of extracellular matrix-degrading enzymes may indicate an important function in production of tumor metastases.

G. The *ski* Oncogene

The nuclear oncogene, v-*ski*, is less well-characterized than those described above. This oncogene was isolated from an avian leukosis virus in culture, and the protein it encodes has several features which suggest it may be a transcription factor.[316] The function of the c-*ski* protooncogene from which it is derived is unknown; however transformation of quail embryo cells with v-*ski* causes them to undergo myogenic differentiation, suggesting a relationship to the MyoD family of myogenic proteins,[196,317] while *ski* transgenic mice have excessive muscle development due to fiber hypertrophy.[318]

H. Tumor Suppressor Genes and Transcription

With the exception of the *WT* gene, the majority of the discussion so far has been concerned with dominant oncogenes, i.e. genes which cause cellular transformation as a result of deregulated expression or mutational activation. The opposing class of genes are the recessive oncogenes or tumor suppressor genes, whose products act to inhibit cell proliferation and promote differentiation. Transformation occurs when the function of both genomic copies of such genes is lost, either by homozygous deletion, mutation, or a combination of the two.

While an increasing number of potential tumor suppressor gene loci has been identified, few are well characterized; however, the products of three of them are localized in the nucleus. One, WT1, has been described already, since it is a member of the ZF protein family, while evidence is accumulating that the others, pRb, and p53, also function as regulators of transcription.

The Retinoblastoma Gene Product

The gene for susceptibility to the embryonic ocular tumor retinoblastoma, *rb*, was the first tumor suppressor gene to be isolated.[319–321] It is homozygously inactivated by deletion or mutation in a high proportion of retinoblastomas and encodes a nuclear phosphoprotein with DNA-binding activity.[321] *rb* was subsequently shown to have a wider role in neoplasia, being inactivated in a proportion of other human cancers.[322–325] In addition, the pRb product is sequestered by various DNA virus tumor antigens, including Adenovirus E1a, SV40 virus large T, and papillomavirus E7 proteins,[326–328] suggesting that disruption of normal pRb function may be advantageous to the process of viral infection.

In normal cells the phosphorylation state of pRb varies with the cell cycle and in nuclear extracts it can be shown to complex with a number of cellular proteins.[329–331] One of these is the transcription factor E2F(DRTF) that transactivates the promoters of several proliferation-related genes.[332–334] During G1 phase of the cell cycle, E2F forms a transcriptionally inactive DNA-binding complex with underphosphorylated pRb.[332] At the G1/S transition phosphorylation of pRb occurs and the complex dissociates. E2F then becomes associated with cyclin A via p107, an Rb-related protein,[335] in a still inactive complex, only becoming functional when it is released by phosphorylation of cyclin A on entry into G2/M. In resting cells, E2F DNA-binding activity is not detected owing to binding of the pRb/E2F by a putative additional factor.[332] By complexing with pRb and p107, the DNA-viral oncoproteins liberate E2F in its active form.[336]

pRb represses the promoters of genes involved in proliferation, such as c-*fos* and c-*myc* which contain potential binding sites for E2F.[337,338] In both of these promoters the E2F site is necessary for pRb-mediated repression. This is not the sole way in which pRb influences transcription, for in addition to repressing transcription of the c-*myc* promoter, the c-Myc protein is among the nuclear factors which complex with pRb as detected by affinity chromatography.[338] Furthermore, besides inactivating growth stimulating proteins, pRb activates transcription of the gene for the growth inhibitory factor TGF-β2. This activation is dependent upon an ATF site in the promoter which preferentially binds the LZ protein ATF-2, while an ATF-2-GAL4 fusion protein confers pRb inducibility upon a promoter containing a GAL4-binding site. Moreover, interaction between ATF-2 and pRb has been detected *in vitro* using a pRb column.[339] Thus it appears that pRb has diverse functions in transcriptional regulation of cell proliferation.

p53

The *p53* gene is probably the most frequent target of spontaneous mutation in human tumors. Its product, like pRb, is a nuclear phosphoprotein with DNA-binding activity,[340] which is sequestered by DNA virus transforming proteins, including Adenovirus E1B, SV40 T, and papillomavirus E6.[341–344] Unlike pRb, the p53 protein exists in cells as dimers or tetramers.

While the normal p53 protein suppresses proliferation,[345–347] mutant forms of p53 have dominant transforming activity, by virtue of their ability to multimerize with the wild-type protein.[348] Structural features of the p53 protein, including its acidic N-terminal domain and basic C-terminal region, are consistent with it being transcriptionally active.[349] Also, a fusion protein of the whole of p53, or its acidic domain with the DNA-binding domain of GAL4, has transactivating activity. In this assay, mutant p53-GAL4 fusion proteins are inactive.[350–352] In addition wild-type p53 represses the transcription of several proliferation-related genes by an unknown mechanism.[353–355]

Two distinct, specific DNA-binding elements for p53 have been identified: one was found in a GC-rich region of the polyoma virus (SV40) enhancer, and the other is a direct repeat of a TGCCT motif separated by a short spacer.[356,357] These elements occur in close proximity to one another, as a p53-responsive 5′ positive enhancer, in the muscle-specific creatine kinase (MCK) gene.[358] Mutant p53 protein fails to activate transcription from this sequence and inhibits transactivation of the polyoma virus promoter by wild-type p53 when the two are coexpressed, probably by preventing DNA-binding of the latter.[359] These findings strongly suggest that the transforming ability of mutant p53 may be due to its interference with the normal transcriptional functions of the wild-type protein.

IV. CONCLUSION

It is clear that transcriptional deregulation as a result of abnormalities in the structure or expression of transcriptionally active proteins plays an important role in the genesis of cancer. This is true both for virally induced tumors in animals and for spontaneous human malignancies. Many aspects of transcription factor function may be affected and the abnormal or overexpressed proteins may have either dominant or recessive oncogenicity. With recent evidence indicating that both p53 and pRb act via transcriptional regulation, it is apparent that transcriptional deregulation is a very common and important feature of human tumors.

A number of unifying themes emerge regarding the mechanisms by which transcription factors are converted into oncogenes. For example, immediate early genes, such as *fos*, *myc*, and *evi-1*, normally expressed only briefly in response to growth stimuli, can become oncogenic as a result of genetic changes which cause loss of normal regulation and constitutive expression. This may result from gene amplification or from translocation or viral transduction leading to expression under the control of a constitutively active promoter. Similar mechanisms may activate developmentally regulated transcription factors, both those expressed during early development of the tissue in which they are oncogenic and those normally active in different tissues. In T-cell leukemias, activation is typically a result of juxtaposition of such a developmental gene to a T-cell receptor gene locus by translocation, while in B-cell tumors the immunoglobulin loci are involved. Other translocations generate chimeric proteins, usually combining a potent transcriptional activation domain from a constitutively active protein, such as that of the bHLH *E2* gene products with a developmentally regulated transcription factor and presumably activating genes normally expressed at an earlier stage of development.

Transcription factors encoded by tumor suppressor genes may be inactivated by deletion or point mutation, allowing uncontrolled cell proliferation. In the case of p53, the mutant protein is a dominant negative oncoprotein which directly interferes with the function of its normal counterpart.

Retroviral oncogenes often exhibit more subtle changes in structure, with a combination of small internal deletions and point mutations, which makes elucidation of the precise mechanism of oncogenic conversion more difficult. For some, such as v-ErbA and v-Rel, the oncoprotein acts in a dominant negative fashion, interfering with the function of endogenous proteins which are responsible for differentiation. Others, including v-Jun and the v-Myb protein produced by AMV, have been rendered oncogenic by the deletion of a transcriptional repressor domain, thereby generating a constitutive transcriptional activator. From studies of viral oncogenes, it is clear that mutations affecting only one or a few amino acids can have drastic effects on transcription factor function. Such changes are not detectable by cytogenetic analysis and may well be important in human malignancies. In addition to the widespread occurrence of point mutations in the *p53* gene, the presence of internal deletions of the *tal-1* gene in 25% of T-ALL and of point mutations in the *WT* gene in a proportion of sporadic Wilms' tumors supports this notion. Although much of the evidence for the involvement of transcription factors, apart from p53 and pRb, in human cancer has come from the leukemias owing to their relative ease of study. It is likely that they will prove equally important in the genesis of solid tumors.

The identification of target genes regulated by the transcription factors involved in neoplasia should provide an improved understanding of aberrant phenotype of tumor cells. Furthermore, detection of the abnormal transcription factors may aid both in the pathological diagnosis of specific tumors and in the assessment of prognosis. For example, immunohistochemical detection of p53 protein, which is only demonstrable in tissue sections when abnormal, may be helpful in mesothelial proliferations in differentiating reactive hyperplasia from malignant mesothelioma,[360] while in breast carcinoma detectable p53 expression is associated with early relapse and shorter survival.[361] In contrast, detection of oestrogen and progesterone receptor expression in breast carcinoma cells indicates both a good prognosis and predicts a response to hormonal manipulation therapy.

In the future, demonstration of tissue-specific transcription factors may provide a more reliable means of determining the cell lineage of a poorly-differentiated tumor and thus lead to implementation of the most appropriate treatment. This may be particularly relevant to soft tissue tumors, in which the tumor cell type may be difficult to determine without resort to electron microscopy. Demonstration of, for example, muscle-specific factors such as MyoD in tumor cells, which do not show expression of muscle structural proteins, could be helpful in reaching a diagnosis and other factors might indicate neural, adipocytic, or vascular origin.

Finally, a detailed understanding of transcription factor structure and function and their role in cancer may enable the development of drugs targeted against specific factors (other than the steroid hormone receptor family, which are already frequent targets of therapy in a range of tumors), thereby interfering with vital steps in the pathways leading to uncoordinated proliferation. In all, much will be learned and much is to be gained from the study of transcription factors in cancer.

ACKNOWLEDGMENT

The author would like to thank Dr. Colin Goding of Marie Curie Cancer Research for help and advice in the preparation of this manuscript.

REFERENCES

1. Ptashne, M. Gene regulation by proteins acting nearby and at a distance. *Nature (London)* **1986**, *322*, 697–701.
2. Dynan, W. S.; Tjian, R. Control of eukaryotic messenger RNA synthesis by sequence-specific DNA-binding proteins. *Nature (London)* **1985**, *316*, 774–778.
3. Maniatis, T.; Goodbourn, S.; Fischer, J. A. Regulation of inducible and tissue-specific gene expression. *Science* **1987**, *236*, 1237–1245.
4. Sawadogo, M.; Sentenac, A. RNA polymerase B (II) and general transcription factors. *Annu. Rev. Biochem.* **1990**, *59*, 711–754.
5. Young, R. A. RNA polymerase II. *Annu. Rev. Biochem.* **1991**, *60*, 689–715.
6. Allison, L. A.; Moyle, M.; Shales, M.; Ingles, C. J. Extensive homology among the largest subunits of eukaryotic and prokaryotic RNA polymerases. *Cell* **1985**, *42*, 599–610.

7. Sentenac, A. Eukaryotic RNA polymerases. *CRC Crit. Rev. Biochem.* **1985**, *18*, 31–90.

8. Ahearn, J. M.; Bartolomei, M. S.; West, M. L.; Cisek, L. J.; Corden, J. L. Cloning and sequence analysis of the mouse genomic locus encoding the largest subunit of RNA polymerase II. *J. Biol. Chem.* **1987**, *262*, 10695–10705.

9. Memet, S.; Gouy, M.; Marck, C.; Sentenac, A.; Buhler, J. M. RPA190, the gene coding for the largest subunit of yeast RNA polymerase A. *J. Biol. Chem.* **1988**, *263*, 2830–2839.

10. Jokerst, R. S.; Weeks, J. R.; Zehring, W. A.; Greenleaf, A. L. Analysis of the gene encoding the largest subunit of RNA polymerase II in *Drosophila. Mol. Gen. Genet.* **1989**, *215*, 266–275.

11. Falkenburg, D.; Dworniczak, B.; Faust, D. M.; Bautz, E. K. F. RNA polymerase II of *Drosophila.* Relation of its 140,000 Mr subunit to the beta subunit of *Escherichia coli* RNA polymerase. *J. Mol. Biol.* **1987**, *195*, 929–937.

12. Sweetser, D.; Nonet, M.; Young, R. A. Prokaryotic and eukaryotic RNA polymerases have homologous core subunits. *Proc. Natl. Acad. Sci. USA* **1987**, *84*, 1192–1196.

13. Gundelfinger, E. D. Interaction of nucleic acids with the DNA-dependent RNA polymerases of *Drosophila. FEBS Lett.* **1983**, *157*, 133–138.

14. Horikoshi, M.; Tamura, H. O.; Sekimizu, K.; Obinata, M.; Natori, S. Identification of the DNA binding subunit of RNA polymerase II from Ehrlich ascites tumor cells. *J. Biochem.* **1983**, *94*, 1761–1767.

15. Bartholomew, B.; Dahmus, M. E.; Meares, C. F. RNA contacts subunits IIo and IIc in HeLa RNA polymerase II transcription complexes. *J. Biol. Chem.* **1986**, *261*, 14226–14231.

16. Chuang, R. Y.; Chuang, L. F. The 180 KDa polypeptide contains the DNA-binding domain of RNA polymerase II. *Biochem. Biophys. Res. Commun.* **1987**, *145*, 73–80.

17. Vilamitjana, J.; Barreau, C. A monoclonal antibody directed against a small subunit of RNA polymerase B blocks the initiation step. *Eur. J. Biochem.* **1987**, *162*, 317–323.

18. Greenleaf, A. L. Amanitin-resistant RNA polymerase II mutations are in the enzyme's largest subunit. *J. Biol. Chem.* **1983**, *258*, 13403–13406.

19. Coulter, D. E.; Greenleaf, A. L. A mutation in the largest subunit of RNA polymerase II alters RNA chain elongation *in vitro. J. Biol. Chem.* **1985**, *260*, 13190–13198.

20. Sigler, P. B. Transcriptional activation. Acid blobs and negative noodles. *Nature (London)* **1988**, *333*, 210–212.

21. Laybourn, P. J.; Dahmus, M. E. Transcription-dependent structural changes in the C-terminal domain of mammalian RNA polymerase subunit IIa/o. *J. Biol. Chem.* **1989**, *264*, 6693–6698.

22. Cisek, L. J.; Corden, J. L. Phosphorylation of RNA polymerase by the murine homologue of the cell-cycle control protein cdc2. *Nature (London)* **1989**, *339*, 679–684.

23. Lee, J. M.; Greenleaf, A. L. A protein kinase that phosphorylates the C-terminal repeat domain of the largest subunit of RNA polymerase II. *Proc. Natl. Acad. Sci. USA* **1989**, *86*, 3624–3628.

24. Stevens, A.; Maupin, M. K. 5,6-Dichloro-1-beta-D-ribofuranosylbenzimidazole inhibits a HeLa protein kinase that phosphorylates an RNA polymerase II-derived peptide. *Biochem. Biophys. Res. Commun.* **1989**, *159*, 508–515.

25. Fischer, L.; Gerard, M.; Chalut, C. Cloning of the 62 kilodalton component of basic transcription factor BTF2. *Science* **1992**, *257*, 1392–1395.

26. Peterson, M. G.; Tjian, R. The tell-tail trigger. *Nature (London)* **1992**, *358*, 620–621.

27. Matsui, T.; Segall, J.; Weil, P. A.; Roeder, R. G. Multiple factors required for accurate initiation of transcription by purified RNA polymerase II. *J. Biol. Chem.* **1980**, *255*, 11992–11996.

28. Dignam, J. D.; Lebovitz, R. M.; Roeder, R. G. Accurate transcription initiation by RNA polymerase II in a soluble extract from isolated mammalian nuclei. *Nucleic Acids Res.* **1983**, *11*, 1475–1489.

29. van Dyke, M. W.; Roeder, R. G.; Sawadogo, M. Physical analysis of transcription preinitiation complex assembly on a class II gene promoter. *Science* **1983**, *241*, 1335–1338.

30. Buratowski, S.; Hahn, S.; Guarente, L.; Sharp, P. A. Five intermediate complexes in transcription initiation by RNA polymerase II. *Cell* **1989**, *56*, 549–561.

31. Breathnach, R.; Chambon, P. Organization and expression of eucaryotic split genes coding for proteins. *Ann. Rev. Biochem.* **1981**, *50*, 349–383.

32. Cavallini, B.; Faus, I.; Matthes, H.; et al. Cloning of the gene encoding the yeast protein BTF1Y, which can substitute for the human TATA box-binding factor. *Proc. Natl. Acad. Sci. USA* **1989**, *86*, 9803–9807.

33. Hahn, S.; Buratowski, S.; Sharp, P. A.; Guarente, L. Isolation of the gene encoding the yeast TATA binding protein TFIID: a gene identical to the SPT15 suppressor of Ty element insertions. *Cell* **1989**, *58*, 1173–1181.

34. Horikoshi, M.; Wang, C. K.; Fujii, H.; Cromlish, J. A.; Weil, P. A.; Roeder, R. G. Cloning and structure of a yeast gene encoding a general transcription initiation factor TFIID that binds to the TATA box. *Nature (London)* **1989**, *341*, 299–303.

35. Schmidt, M. C.; Kao, C. C.; Pei, R.; Berk, A. J. Yeast TATA-box transcription factor gene. *Proc. Natl. Acad. Sci. USA* **1989**, *86*, 7785–7789.

36. Kao, C. C.; Lieberman, P. M.; Schmidt, M. C.; Zhou, Q.; Pei, R.; Berk, A. J. Cloning of a transcriptionally active human TATA binding factor. *Science* **1990**, *248*, 1646–1650.

37. Peterson, M. G.; Tanese, M.; Pugh, B. F.; Tjian, R. Functional domains and upstream activation properties of cloned human TATA binding protein. *Science* **1990**, *248*, 1625–1630.

38. Dynlacht, B. D.; Hoey, T.; Tjian, R. Isolation of coactivators associated with the TATA-binding protein that mediate transcriptional activation. *Cell* **1991**, *66*, 563–576.

39. Meisterernst, M.; Roeder, R. G. Family of proteins that interact with TFIID and regulate promoter activity. *Cell* **1991**, *67*, 557–567.

40. Simon, M. C.; Fisch, T. M.; Benecke, B. J.; Nevins, J. R.; Heintz, N. Definition of multiple, functionally distinct TATA elements, one of which is a target in the hsp70 promoter for E1A regulation. *Cell* **1988**, *52*, 723–729.

41. Wefald, F. C.; Devlin, B. H.; Williams, R. S. Functional heterogeneity of mammalian TATA-box sequences revealed by interaction with a cell-specific enhancer. *Nature (London)* **1990**, *344*, 260–262.

42. Pugh, B. F.; Tjian, R. Transcription from a TATA-less promoter requires a multisubunit TFIID complex. *Genes Dev.* **1991**, *5*, 1935–1945.

43. Comai, L.; Tanese, N.; Tjian, R. The TATA-binding protein and associated factors are integral components of the RNA polymerase I transcription factor, SL1. *Cell* **1992**, *68*, 965–976.

44. White, R. J.; Jackson, S. P.; Rigby, P. W. J. A role for the TATA-box-binding protein component of the transcription factor IID complex as a general RNA polymerase III transcription factor. *Proc. Natl. Acad. Sci. USA* **1992**, *89*, 1949–1953.

45. Sharp, P. A. TATA-binding protein is a classless factor. *Cell* **1992**, *68*, 819–821.

46. White, R. J.; Jackson, S. P. The TATA-binding protein: a central role in transcription by RNA polymerases I,II and III. *Trends Genet.* **1992**, *8*, 284–288.

47. Davison, B. L.; Egly, J. M.; Mulvill, E. R.; Chambon, P. Formation of stable preinitiation complexes between eukaryotic class B transcription factors and promoter sequences. *Nature (London)* **1983**, *301*, 680–686.

48. Reinberg, D.; Horikoshi, M.; Roeder, R. G. Factors involved in specific transcription in mammalian RNA polymerase II. Functional analysis of initiation factors IIA and IID and identification of a new factor operating at sequences downstream of the initiation site. *J. Biol. Chem.* **1987**, *262*, 3322–3330.

49. Ha, I.; Lane, W. S.; Reinberg, D. Cloning of a human gene encoding the general transcription initiation factor IIB. *Nature (London)* **1991**, *352*, 689–695.

50. Cai, H.; Luse, D. S. Transcription initiation by RNA polymerase II *in vitro*. Properties of preinitiation, initiation, and elongation complexes. *J. Biol. Chem.* **1987**, *262*, 298–304.

51. Carthew, R. W.; Samuels, M.; Sharp, P. A. Formation of transcription preinitiation complexes with an amanitin-resistant RNA polymerase II. *J. Biol. Chem.* **1988**, *263*, 17128–17135.

52. Bunick, D.; Zandomeni, R.; Ackerman, S.; Weinmann, R. Interaction of nucleic acids with the DNA-dependent RNA polymerases of *Drosophila*. *Cell* **1982**, *29*, 877–886.

53. Sawadogo, M.; Roeder, R. G. Energy requirement for specific transcription initiation by the human RNA polymerase II system. *J. Biol. Chem.* **1984**, *259*, 5321–5326.

54. Sopta, M.; Burton, Z. F.; Greenblatt, J. Structure and associated DNA-helicase activity of a general transcription initiation factor that binds to RNA polymerase II. *Nature (London)* **1989**, *341*, 410–414.

55. Hawley, D. K.; Roeder, R. G. Separation and partial characterization of three functional steps in transcription initiation by human RNA polymerase II. *J. Biol. Chem.* **1985**, *260*, 8163–8172.

56. Sekimizu, K.; Yokoi, H.; Natori, S. Evidence that stimulatory factor(s) of RNA polymerase II participates in accurate transcription in a HeLa cell lysate. *J. Biol. Chem.* **1982**, *257*, 2719–2721.

57. Flores, O.; Maldonado, E.; Reinberg, D. Factors involved in specific transcription by mammalian RNA polymerase II. Factors IIE and IIF independently interact with RNA polymerase II. *J. Biol. Chem.* **1989**, *264*, 8913–8921.

58. Danielsen, M.; Hinck, L.; Ringold, G. M. Two amino acids within the knuckle of the first zinc finger specify DNA element activation by the glucocorticoid receptor. *Cell* **1989**, *57*, 1131–1138.

59. Hanes, S. D.; Brent, R. DNA specificity of the bicoid activator protein is determined by homeodomain recognition helix residue 9. *Cell* **1989**, *57*, 1275–1283.

60. Mader, S.; Kumar, V.; de Verneuil, H.; Chambon, P. Three amino acids of the oestrogen receptor are essential to its ability to distinguish an oestrogen from a glucocorticoid-responsive element. *Cell* **1989**, *57*, 1139–1146.

61. Umesono, K.; Evans, R. M. Determinants of target gene specificity for steroid/thyroid hormone receptors. *Cell* **1989**, *57*, 1139–1146.

62. Fisher, F.; Jayaraman, P.-S.; Goding, C. R. C-Myc and the yeast transcription factor PHO4 share a common DNA-binding motif. *Oncogene* **1991**, *6*, 1099–1104.

63. Fisher, F.; Goding, C. R. *EMBO J.* **1992**. In press.

64. Halazonetis, T. D.; Kandil, A. N. Determination of the c-MYC DNA-binding site. *Proc. Natl. Acad. Sci. USA* **1991**, *88*, 6162–6166.

65. Dang, C. V.; Dolde, C.; Gillison, M. L.; Kato, G. J. Discrimination between related DNA sites by a single amino acid residue of Myc-related basic-helix-loop-helix proteins. *Proc. Natl. Acad. Sci. USA* **1992**, *89*, 599–602.

66. Murre, C.; McCaw, P. S.; Baltimore, D. A new DNA-binding and dimerization motif in immunoglobulin enhancer-binding, daughterless, MyoD and Myc proteins. *Cell* **1989**, *56*, 777–783.

67. Ogawa, N.; Oshima, Y. Functional domains of a positive regulatory protein, PHO4, for transcriptional control of the phosphatase regulon in *Saccharomyces cerevisiae*. *Mol. Cell. Biol.* **1990**, *10*, 2224–2236.

68. Villares, R.; Cabrera, C. V. The achaete-scute gene complex of D. melanogaster: conserved domains in a subset of genes required for neurogenesis and their homology to *myc*. *Cell* **1987**, *50*, 415–424.

69. Kouzarides, T.; Ziff, E. Leucine zippers of fos, jun and GCN4 dictate dimerization specificity and thereby control DNA binding. *Nature (London)* **1989**, *340*, 568–571.

70. Vinson, C. R.; Sigler, P. B.; McKnight, S. L. Scissors grip model for DNA recognition by a family of leucine zipper proteins. *Science* **1989**, *246*, 911–916.

71. Smeal, T.; Angel, P.; Meek, J.; Karin, M. Different requirements for formation of *jun:jun* and *jun:fos* complexes. *Genes Dev.* **1989**, *3*, 2091–2100.

72. Hai, T.; Liu, F.; Coukos, W. J.; Green, M. R. Transcription factor ATF cDNA clones: an extensive family of leucine zipper proteins able to selectively form DNA-binding heterodimers. *Genes Dev.* **1989**, *3*, 2083–2090.

73. Hu, Y.-F.; Luscher, B.; Admon, A.; Mermod, N.; Tjian, R. Transcription factor AP4 contains multiple dimerization domains that regulate dimer specificity. *Genes Dev.* **1990**, *4*, 1741–1752.

74. Müller, M. M.; Ruppert, S.; Schaffner, W.; Matthias, P. A cloned octamer transcription factor stimulates transcription from lymphoid-specific promoters in non-B cells. *Nature (London)* **1988**, *336*, 544–551.

75. Miller, J.; McLachlan, A. D.; Klug, A. Repetitive zinc-binding domains in the protein transcription factor TFIIIA from *Xenopus* oocytes. *EMBO J.* **1985**, *4*, 1609–1614.

76. Brown, R. S.; Sander, C.; Argos, P. The primary structure of transcription factor TFIIIA has 12 consecutive repeats. *FEBS Lett.* **1985**, *186*, 271–274.

77. Page, D. C.; Mosher, R.; Simpson, E. M.; et al. The sex determining region of the human Y chromosome encodes a finger protein. *Cell* **1987**, *51*, 1091–1094.

78. Diakun, G. P.; Fairall, L.; Klug, A. EXAFS study of the zinc-binding sites in the protein transcription factor IIIA. *Nature (London)* **1986**, *324*, 698–699.

79. Kadonaga, J. T.; Carner, K. R.; Masiarz, F. R.; Tjian, R. Isolation of cDNA encoding transcription factor SP1 and functional analysis of the DNA binding domain. *Cell* **1987**, *51*, 1079–1090.

80. Green, S.; Chambon, P. Oestradiol induction of a glucocorticoid-responsive gene by a chimeric receptor. *Nature (London)* **1987**, *325*, 75–78.

81. Hard, T.; Kellenbach, E.; Boelens, R.; et al. Solution structure of the glucocorticoid receptor DNA-binding domain. *Science* **1990**, *249*, 157–160.

82. Vallee, B. L.; Coleman, J. E.; Auld, D. S. Zinc fingers, zinc clusters, and zinc twists in DNA-binding protein domains. *Proc. Natl. Acad. Sci. USA* **1991**, *88*, 999–1003.

83. McGinnis, W.; Levine, M. S.; Hafen, E.; Kuroiwa, A.; Gehring, W. J. A conserved DNA sequence in homoeotic genes of the *Drosophila, Antennapedia*, and bithorax complexes. *Nature (London)* **1984**, *308*, 428–433.

84. Scott, M. P.; Weiner, A. J. Structural relationships among genes that control development: sequence homology between the *Antennapedia*, ultrabithorax, and *fushi tarazu* loci of *Drosophila*. *Proc. Natl. Acad. Sci. USA* **1984**, *81*, 4115–4119.

85. Gehring, W. J.; Hiromi, Y. Homeotic genes and the homeobox. *Ann. Rev. Genet.* **1986**, *20*, 147–173.

86. Gehring, W. J. Homeoboxes in the study of development. *Science* **1987**, *236*, 1245–1252.

87. Hoey, T.; Levine, M. Divergent homeo box proteins recognize similar DNA sequences in *Drosophila*. *Nature(London)* **1988**, *332*, 858–861.

88. Kuziora, M. A.; McGinnis, W. A homeodomain substitution changes the regulatory specificity of the deformed protein in *Drosophila* embryos. *Cell* **1989**, *59*, 563–571.

89. Laughon, A.; Scott, M. P. Sequence of a *Drosophila* segmentation gene: protein structure homology with DNA-binding proteins. *Nature (London)* **1984**, *310*, 25–31.

90. Shepherd, J. C.; McGinnis, W.; Carrasco, A. E.; et al. Fly and frog homeodomains show homology with yeast mating type regulatory proteins. *Nature (London)* **1984**, *310*, 70–71.

91. Pabo, C. O.; Sauer, R. T. Protein-DNA recognition. *Ann. Rev. Biochem.* **1984**, *53*, 293–321.

92. Kissinger, C. R.; Liu, B.; Martin-Blanco, E.; Kornberg, T. B.; Pabo, C.O. Crystal structure of an engrailed homeodomain-DNA complex at 2.8 Å resolution: A framework for understanding homeodomain-DNA interactions. *Cell* **1990**, *63*, 579–590.

93. Sturm, R. A.; Das, G.; Herr, W. The ubiquitous octamer-binding protein Oct-1 contains a POU domain with a homeo box subdomain. *Genes Dev.* **1988**, *2*, 1582–1599.

94. Desplan, C.; Theis, J.; O'Farrell, P. H. The sequence specificity of homeodomain-DNA interaction. *Cell* **1988**, *54*, 1081–1090.

95. Sturm, R. A.; Herr, W. The POU domain is a bipartite DNA-binding structure. *Nature (London)* **1988**, *336*, 601–604.

96. Ingraham, H. A.; Flynn, S. E.; Voss, J. W.; et al. The POU-specific domain of Pit-1 is essential for sequence-specific, high affinity DNA binding and DNA-dependent Pit-1-Pit-1 interactions. *Cell* **1990**, *61*, 1021–1033.

97. Clerc, R. G.; Corcoran, L. M.; LeBowitz, J. H.; Baltimore, D.; Sharp, P. A. The B-cell-specific Oct-2 protein contains POU box- and homeo box-type domains. *Genes Dev.* **1988**, *2*, 1570–1581.

98. Mitchell, P. J.; Tjian, R. Transcriptional regulation in mammalian cells by sequence-specific DNA binding proteins. *Science* **1989**, *245*, 371–378.

99. Ptashne, M.; Gann, A. A. F. Activators and targets. *Nature (London)* **1990**, *346*, 329–331.

100. Stringer, K. F.; Ingles, C. J.; Greenblatt, J. Direct and selective binding of an acidic transcriptional activation domain to the TATA-box factor TFIID. *Nature (London)* **1990**, *345*, 783–786.

101. Lin, Y.-S.; Green, M. R. Mechanism of action of an acidic transcriptional activator *in vitro*. *Cell* **1991**, *64*, 971–981.

102. Hoey, T.; Dynlacht, B. D.; Peterson, M. G.; Pugh, B. F.; Tjian, R. Isolation and characterization of the *Drosophila* gene encoding the TATA box binding protein, TFIID. *Cell* **1990**, *61*, 1179–1186.

103. Pugh, B. F.; Tjian, R. Mechanism of transcriptional activation by Sp1: Evidence for coactivators. *Cell* **1990**, *61*, 1187–1197.

104. Bodner, M.; Karin, M. A pituitary-specific trans-acting factor can stimulate transcription from the growth hormone promoter in extracts of nonexpressing cells. *Cell* **1987**, *50*, 267–275.

105. Ingraham, H. A.; Chan, R.; Mangalam, H. J.; et al. A tissue-specific transcription factor containing a homeodomain specifies a pituitary phenotype. *Cell* **1988**, *55*, 519–529.

106. Mangalam, H. J.; Albert, V. R.; Ingraham, H. A.; et al. A pituitary POU domain protein, Pit-1, activates both growth hormone and prolactin promoters transcriptionally. *Genes Dev.* **1989**, *3*, 946–958.

107. Schuster, W. A.; Treacy, M. N.; Martin, F. Tissue specific *trans*-acting factor interaction with proximal rat prolactin gene promoter sequences. *EMBO J.* **1988**, *7*, 1721–1733.

108. Nelson, C.; Albert, V. R.; Elsholtz, H. P.; Lu, L. I.-W.; Rosenfeld, M. G. Activation of cell-specific expression of rat growth hormone and prolactin genes by a common transcription factor. *Science* **1988**, *239*, 1400–1405.

109. Dana, S.; Karin, M. Induction of human growth hormone promoter activity by the adenosine 3′,5′-monophosphate pathway involves a novel responsive element. *Mol. Endocrinol.* **1989**, *3*, 815–821.

110. Copp, R. P.; Samuels, H. H. Identification of an adenosine 3′,5′-monophosphate (cAMP)-responsive region in the rat growth hormone gene: Evidence for independent and synergistic effects of cAMP and thyroid hormone on gene expression. *Mol. Endocrinol.* **1989**, *3*, 790–796.

111. Johnson, P. F.; McKnight, S. L. Eukaryotic transcriptional regulatory proteins. *Ann. Rev. Biochem.* **1989**, *58*, 799–839.

112. Dorn, A.; Bollekens, J.; Staub, A.; Benoist, C.; Mathis, D. A multiplicity of CCAAT box-binding proteins. *Cell* **1987**, *50*, 863–872.

113. Chodosh, L. A.; Baldwin, A. S.; Carthew, R. W.; Sharp, P. A. Human CCAAT-binding proteins have heterologous subunits. *Cell* **1988**, *53*, 11–17.

114. Raymondjean, M.; Cereghini, S.; Yaniv, M. Several distinct "CCAAT" box binding proteins coexist in eukaryotic cells. *Proc. Natl. Acad. Sci. USA* **1988**, *85*, 757–761.

115. Cox, P. M.; Temperley, S. M.; Kumar, H.; Goding, C. R. A distinct octamer binding protein present in malignant melanoma cells. *Nucleic Acids Res.* **1988**, *16*, 11047–11056.

116. McKnight, S. L.; Tjian, R. Transcriptional selectivity of viral genes in mammalian cells. *Cell* **1986**, *46*, 795–805.

116a. Smeal, T.; Angel, P.; Meek, J.; Karin, M. Different requirements for formation of *jun:jun* and *jun:fos* complexes. *Genes Dev.* **1989**, *3*, 2091–2100.

117. Benbrook, D. M.; Jones, N. C. Heterodimer formation between Creb and Jun proteins. *Oncogene* **1990**, *5*, 295–302.

118. MacDonald, N. J.; Kuhl, D.; Maguire, D.; et al. Different pathways mediate virus-inducibility of the human *IFN*–α1 and *IFN*–β genes. *Cell* **1990**, *60*, 767–779.

119. Miyamoto, M.; Fujita, T.; Kimura, Y.; et al. Regulated expression of a gene encoding a nuclear factor, IFR-1, that specifically binds to *IFN*–β gene regulatory elements. *Cell* **1988**, *54*, 903–913.

120. Harada, H.; Fujita, T.; Miyamoto, M.; et al. Structurally similar but functionally distinct factors, *IRF*–1 and *IRF*-2, bind to the same regulatory elements of *IFN* and *IFN*-inducible genes. *Cell* **1989**, *58*, 729–739.

121. Hunter, T.; Karin, M. The regulation of transcription by phosphorylation. *Cell* **1992**, *70*, 375–387.

122. Moll, T.; Tebb, G.; Surana, U.; Robitsch, H.; Nasmyth, K. The role of phosphorylation and the CDC28 protein kinase in cell cycle-regulated nuclear import of the *S. cerevisiae* transcription factor SWI5. *Cell* **1991**, *66*, 743–758.

123. Gilmore, T. D. Malignant transformation by mutant Rel proteins. *Trends Genet.* **1991**, *7*, 1–5.

124. Lüscher, B.; Christenson, E.; Litchfield, D. W.; Krebs, E. G.; Eisenman, R. N. Myb DNA binding is inhibited by phosphorylation at a site deleted during oncogenic activation. *Nature* **1990**, *344*, 517–522.

125. Boyle, W. J.; Smeal, T.; Defize, L. H. K.; et al. Activation of protein kinase C decreases phosphorylation of c-*jun* at sites that negatively regulate its DNA-binding activity. *Cell* **1991**, *64*, 573–584.

126. Papavassiliou, A. G.; Bohmann, K.; Bohmann, D. Determining the effect of inducible protein phosphorylation on the DNA-binding activity of transcription factors. *Anal. Biochem.* **1992**, *203*, 302–309.

127. Janknecht, R.; Hipskind, R. A.; Houthaeve, T.; Nordheim, A.; Stunnenberg, H. G. Identification of multiple SRF N-terminal phosphorylation sites affecting DNA-binding properties. *EMBO J.* **1992**, *11*, 1045–1054.

128. Marais, R. M.; Hsuan, J. J.; McGuigan, C.; Wynne, J.; Treisman, R. Casein kinase II phosphorylation increases the rate of serum response factor-binding site exchange. *EMBO J.* **1992**, *11*, 97–105.

129. Cherry, J. R.; Johnson, T. R.; Dollard, C. A.; Shuster, J. R.; Denis, C. L. Cyclic AMP-dependent protein kinase phosphorylates and inactivates the yeast transcriptional activator ADR1. *Cell* **1989**, *56*, 409–419.

130. Gonzalez, G. A.; Yamamoto, K. K.; Fischer, W. H.; et al. A cluster of phosphorylation sites on the cyclic AMP-regulated nuclear factor CREB predicted by its sequence. *Nature(London)* **1989**, *337*, 749–752.

131. Pulverer, B. J.; Kyriakis, J. M.; Avruch, J.; Nikolakaki, E.; Woodgett, J. R. Phosphorylation of c-*jun* mediated by MAP kinases. *Nature (London)* **1991**, *353*, 670–674.

132. Smeal, T.; Binétruy, B.; Mercola, D.; Birrer, M.; Karin, M. Oncogenic and transcriptional cooperation with Ha-Ras requires phosphorylation of c-*jun* on serines 63 and 73. *Mol. Cell. Biol.* **1991**. In press.

133. Ofir, R.; Dwarki, V. J.; Rashid, D.; Verma, I. M. Phosphorylation of the C-terminus of Fos protein is required for transcriptional transrepression of the c-*fos* promoter. *Nature (London)* **1990**, *348*, 80–82.

134. Baeuerle, P. A.; Baltimore, D. I-κB: a specific inhibitor of the NF-κB transcription factor. *Science* **1988**, *242*, 540–546.

135. Benezra, R.; Davis, R. L.; Lockshon, D.; Turner, D. L.; Weintraub, H. The protein Id : a negative regulator of helix-loop-helix DNA binding proteins. *Cell* **1990**, *61*, 49–59.

136. Ellis, H. M.; Spann, D. R.; Posakony, J. W. Extramacroachaete: a negative regulator of sensory organ development in *Drosophila*, defines a new class of helix-loop-helix proteins. *Cell* **1990**, *61*, 27–38.

137. Garrell, J.; Modolell, J. The *Drosophila* extramacroachaete locus, an antagonist of proneural genes that, like these genes, encodes a helix-loop-helix protein. *Cell* **1990**, *61*, 39–48.

138. Auwerx, J.; Sassone-Corse, P. IP-1: A dominant inhibitor of *fos/jun* whose activity is modulated by phosphorylation. *Cell* **1991**, *64*, 983–993.

139. Treacy, M. N.; He, X.; Rosenfeld, M. G. I-POU: a POU-domain protein that inhibits neuron-specific gene activation. *Nature (London)* **1991**, *350*, 577–584.

140. Beato, M. Gene regulation by steroid hormones. *Cell* **1989**, *56*, 335–344.

141. Green, S.; Chambon, P. Nuclear receptors enhance our understanding of transcription regulation. *Trends Genet.* **1988**, *4*, 309–314.

142. Nüsslein-Volhard, C.; Wieschaus, E. Mutations affecting segment number and polarity in *Drosophila*. *Nature (London)* **1980**, *287*, 795–801.

143. Akam, M. The molecular basis for metameric pattern in the *Drosophila* embryo. *Development* **1987**, *101*, 1–22.

144. Scott, M. P.; Carroll, S. B. The segmentation and homeotic gene network in early *Drosophila* development. *Cell* **1987**, *51*, 689–698.

145. Manley, J. L.; Levine, M. S. The homeobox and mammalian development. *Cell* **1985**, *43*, 1–2.

146. Holland, P. W. H.; Hogan, B. L. M. Expression of homeobox genes during mouse development: A review. *Genes Dev.* **1988**, *2*, 773–782.

147. Graham, A.; Papolopulu, N.; Krumlauf, R. The murine and *Drosophila* homeobox gene complexes have common features of organization and expression. *Cell* **1989**, *57*, 367–378.

148. Finney, M.; Ruvkun, G. The *unc*-86 gene product couples cell lineage and cell identity in *C. elegans*. *Cell* **1990**, *63*, 895–905.

149. Way, J. C.; Chalfie, M. The *mec*-3 gene of *Caenorhabditis elegans* requires its own product for maintained expression and is expressed in three neuronal cell types. *Genes Dev.* **1989**, *3*, 1823–1833.

150. Li, S.; Crenshaw, E. B.; Rawson, E. J.; Simmons, D. M.; Swanson, L. W.; Rosenfeld, M. G. Dwarf locus mutants lacking three pituitary cell types result from mutations in the POU-domain gene *pit*-1. *Nature (London)* **1990**, *347*, 528–533.

151. Martin, G. S. Rous sarcoma viruses: A function required for the maintenance of the transformed state. *Nature* **1970**, *227*, 1021–1023.

152. Varmus, H.; Bishop, J. M. (Eds.) Biochemical mechanisms of oncogene activity: proteins encoded by oncogenes. *Cancer Surv.* **1986**, *5*, 153.

153. Pawson, T. Transcription factors as oncogenes. *Trends Genet.* **1987**, *3*, 333–334.

154. Maki, Y.; Bos, T. J.; Davis, C.; Starbuck, M.; Vogt, P. K. Avian sarcoma virus 17 carries the *jun* oncogene. *Proc. Natl. Acad. Sci. USA* **1987**, *84*, 8248–8252.

155. Vogt, P. K.; Bos, T. J.; Doolittle, R. F. Homology between the DNA binding domain of the GCN4 regulatory protein of yeast and the carboxy terminal region of a protein coded for by the oncogene *jun*. *Proc. Natl. Acad. Sci. USA* **1987**, *84*, 3316–3319.

156. Struhl, K. The Jun oncoprotein, a vertebrate transcription factor, activates transcription in yeast. *Nature (London)* **1988**, *334*, 649–650.

157. Hill, D. E.; Hope, I. A.; Macke, J. P.; Struhl, K. Saturation mutagenesis of the yeast *his 3* regulatory site: requirements for transcriptional induction and for binding by GCN4 activator protein. *Science* **1986**, *234*, 451–457.

158. Lee, W.; Mitchell, P.; Tjian, R. Purified transcription factor AP-1 interacts with TPA-inducible enhancer elements. *Cell* **1987**, *49*, 741–752.

159. Angel, P.; Imagura, M.; Chiu, R.; et al. Phorbol ester-inducible genes contain a common *cis* element recognized by a TPA-modulated transacting factor. *Cell* **1987**, *49*, 729–739.

160. Bos, T. J.; Bohmann, D.; Tsuchie, H.; Tjian, R.; Vogt, P.K. v-*jun* encodes a nuclear protein with enhancer binding properties of AP1. *Cell* **1988**, *52*, 705–712.

161. Angel, P.; Allegretto, E. A.; Okino, S. T.; et al. Oncogene jun encodes a sequence-specific *trans*-activator similar to AP-1. *Nature (London)* **1988**, *332*, 166–171.

162. Bos, T. J.; Monetclaro, F. S.; Mitsunobu, F.; et al. Efficient transformation of chicken embryo fibroblasts by c-*jun* requires structural modification in coding and non-coding sequences. *Genes Dev.* **1990**, *4*, 1677–1687.

163. Baichwal, V. R.; Tjian, R. Control of c-*jun* activity by integration of a cell-specific inhibitor with regulatory domain β: differences between v- and c-*jun*. *Cell* **1990**, *63*, 815–825.

164. Van Beveren, C.; van Straaten, F.; Curran, T.; Muller, R.; Verma, I. M. Analysis of FBJ-MuSV provirus and c-*fos* (mouse) gene reveals that viral and cellular *fos* gene products have different carboxy termini. *Cell* **1983**, *32*, 1241–1255.

165. Curran, T.; Miller, A. D.; Zokas, L.; Verma, I. M. Viral and cellular fos proteins: A comparative analysis. *Cell* **1984**, *36*, 259–268.

166. Distel, R. J.; Ro, H.-S.; Rosen, B. S.; Groves, D. L.; Spiegelman, B. M. Nucleoprotein complexes that regulate gene expression in adipocyte differentiation: Direct participation of c-*fos*. *Cell* **1987**, *49*, 835–844.

167. Sassone-Corsi, P.; Lamph, W. W.; Kamps, M.; Verma, I. M. fos-associated cellular p39 is related to nuclear transcription factor AP-1. *Cell* **1988**, *54*, 553–560.

168. Rauscher, F. J.; Sambucetti, L. C.; Curran, T.; Distel, R. J.; Spiegelman, B. M. Common DNA binding site for Fos protein complexes and transcription factor AP-1. *Cell* **1988**, *52*, 471–480.

169. Chiu, R.; Boyle, W. J.; Meek, J.; Smeal, T.; Hunter, T.; Karin, M. The c-Fos protein interacts with c-*jun*/AP-1 to stimulate transcription of AP-1 responsive genes. *Cell* **1988**, *54*, 541–552.

170. Miller, A. D.; Curran, T.; Verma, I. M. c-Fos protein can induce cellular transformation: A novel mechanism of activation of a cellular oncogene. *Cell* **1984**, *36*, 51–60.

171. Schuermann, M.; Neuberg, M.; Hunter, J. B.; et al. The leucine repeat motif in Fos protein mediates complex formation with Jun/AP1 and is required for transformation. *Cell* **1989**, *56*, 507–516.

172. Wu, J.-X.; Carpenter, P. M.; Gresens, C.; et al. The protooncogene c-*fos* is over-expressed in the majority of human osteosarcomas. *Oncogene* **1990**, *5*, 989–1000.

173. Nagai, M. A.; Habr-Gama, A.; Oshima, C. T.; Brentani, M. M. Association of genetic alterations of c-*myc*, c-*fos*, and c-Ha-*ras* protooncogenes in colorectal tumours. *Dis. Colon Rectum* **1992**, *35*, 444–451.

174. Ryder, K.; Lau, L. F.; Nathans, D. A gene activated by growth factors is related to the oncogene v-*jun*. *Proc. Natl. Acad. Sci. USA* **1988**, *85*, 1487–1491.

175. Ryder, K.; Lanahan, A.; Perez-Albuerne, E.; Nathans, D. *jun* D: A third member of the *jun* gene family. *Proc. Natl. Acad. Sci. USA* **1989**, *86*, 1500–1503.

176. Hirai, S.-I.; Ryseck, R.-P.; Mechta, F.; Bravo, R.; Yaniv, M. Characterization of *jun* D: a new member of the *jun* protooncogene family. *EMBO J.* **1989**, *8*, 1433–1439.

177. Zerial, M.; Toschi, L.; Ryseck, R.-P.; Schuermann, M.; Muller, R.; Bravo, R. The product of a novel growth factor activated gene, fosB, interacts with Jun proteins enhancing their DNA binding activity. *EMBO J.* **1989**, *8*, 805–813.

178. Cohen, D. M.; Curran, T. *fra-1*: A serum-inducible, cellular immediate-early gene that encodes a *fos*-related antigen. *Mol. Cell. Biol.* **1988**, *8*, 2063–2069.

179. Schüle, R.; Evans, R. M. Cross-coupling of signal transduction pathways: Zinc finger meets leucine zipper. *Trends Genet.* **1991**, *7*, 377–381.

180. Diamond, M. I.; Miner, J. N.; Yoshinaga, S. K.; Yamamoto, K. R. Transcription factor interactions: Selectors of positive or negative regulation from a single DNA element. *Science* **1990**, *249*, 1266–1272.

181. Schüle, R.; Umesono, K.; Mangelsdorf, D. J.; Bolado, J.; Pike, J. W.; Evans, R. M. *jun-fos* and receptors for vitamins A and D recognize a common response element in the human osteocalcin gene. *Cell* **1990**, *61*, 497–504.

182. Jonat, G.; Rahmsdorf, H. J.; Park, K.-K.; et al. Antitumor promotion and antiinflammation: down-modulation of AP-1 (*fos/jun*) activity by glucocorticoid hormone. *Cell* **1990**, *62*, 1189–1204.

183. Schüle, R.; Rangarajan, P.; Kliewer, S.; et al. Functional antagonism between oncoprotein c-*jun* and the glucocorticoid receptor. *Cell* **1990**, *62*, 1217–1226.

184. Yang-Yen, H.-F.; Chambard, J.-C.; Sun, Y.-L.; et al. Transcriptional interference between c-*jun* and the glucocorticoid receptor: Mutual interference of DNA binding due to direct protein-protein interaction. *Cell* **1990**, *62*, 1205–1215.

185. Nishizawa, M.; Kataoka, K.; Goto, N.; Fujiwara, K. T.; Kawai, S. v-*maf*, a viral oncogene that encodes a "leucine zipper" motif. *Proc. Natl. Acad. Sci. USA* **1989**, *86*, 7711–7715.

186. Inaba, T.; Roberts, W. M.; Shapiro, L. H.; et al. Fusion of the leucine zipper gene *HLF* to the *E2A* gene in human acute B-lineage leukaemia. *Science* **1992**, *257*, 531–534.

187. Abrams, H. D.; Rohrschneider, L. R.; Eisenman, R. N. Nuclear location of the putative transforming protein of avian myelocytomatosis virus. *Cell* **1982**, *29*, 427–439.

188. Donner, P.; Greiser-Wilke, I.; Moelling, K. Nuclear localization and DNA binding of the transforming gene product of avian myelocytomatosis virus. *Nature* **1982**, *296*, 262–266.

189. Kelly, K.; Cochran, B. H.; Stiles, C. D.; Leder, P. Cell-specific regulation of the c-*myc* gene by lymphocyte mitogens and platelet-derived growth factor. *Cell* **1983**, *35*, 603–610.

190. Erikson, J.; Ar-Rushdi, A.; Drwinga, H. L.; Nowell, P. C.; Croce, C. M. Transcriptional activation of the translocated c-*myc* oncogene in Burkitt lymphoma. *Proc. Natl. Acad. Sci. USA* **1983**, *80*, 820–824.

191. Cole, M. D. The *myc* oncogene: Its role in transformation and differentiation. *Ann. Rev. Genet* **1986**, *20*, 361–384.

192. Nau, M. M.; Brooks, B. J.; Battey, J.; et al. L-*myc*, a new *myc*-related gene amplified and expressed in human small cell lung cancer. *Nature* **1985**, *318*, 69–73.

193. Schwab, M.; Alitalo, K.; Klempnauer, K.-H.; et al. Amplified DNA with limited homology to *myc* cellular oncogene is shared by human neuroblastoma cell lines and a neuroblastoma tumour. *Nature (London)* **1983**, *305*, 245–248.

194. Land, H.; Parada, L. F.; Weinberg, R. A. Tumorigenic conversion of primary embryo fibroblasts requires at least two cooperating oncogenes. *Nature* **1983**, *304*, 596–602.

195. Davis, R. L.; Cheng, P.-F.; Lassar, A. B.; Weintraub, H. The MyoD DNA-binding domain contains a recognition code for muscle-specific gene activation. *Cell* **1990**, *60*, 733–746.

196. Murre, C.; McCaw, P. S.; Vaessin, H.; et al. Interactions between heterologous helix-loop-helix proteins generate complexes that bind specifically to a common DNA sequence. *Cell*, **1989**, *58*, 537–544.

197. Dang, C. V.; McGuire, M.; Buckmire, M.; Lee, W. M. F. Involvement of the leucine zipper region in the oligomerization and transforming activity of the c-Myc protein. *Nature (London)* **1989**, *337*, 664–666.

198. Blackwell, T. K.; Kretzner, L.; Blackwood, E. M.; Eisenman, R.; Weintraub, H. Sequence-specific DNA binding by the c-Myc protein. *Science* **1990**, *250*, 1149–1151.

199. Blackwood, E. M.; Eisenman, R. N. Max: A helix-loop-helix zipper protein that forms a sequence-specific DNA-binding complex with Myc. *Science* **1991**, *251*, 1211–1217.

200. Blackwood, E. M.; Luscher, B.; Eisenman, R. N. Myc and Max associate *in vivo*. *Genes Dev.* **1991**, *6*, 71–80.

201. Kato, G. J.; Lee, W. M.; Chen, L. L.; Dang, C. V. Max: Functional domains and interaction with c-Myc. *Genes Dev.* **1992**, *6*, 81–92.

202. Cole, M. D. Myc meets its Max. *Cell* **1991**, *65*, 715–716.

203. Kretzner, L.; Blackwood, E. M.; Eisenman, R. N. Myc and Max proteins possess distinct transcriptional activities. *Nature (London)* **1992**, *359*, 426–429.

204. Berberich, S. J.; Cole, M. D. Casein kinase II inhibits the DNA-binding activity of Max homodimers but not Myc/Max heterodimers. *Genes Dev.* **1992**, *6*, 166–176.

205. Kato, G. J.; Barrett, J.; Villa-Garcia, M.; Dang, C. V. The amino terminal c-Myc domain required for neoplastic transformation activates transcription. *Mol. Cell. Biol.* **1990**, *10*, 5914–5920.

206. Stone, J.; de Lange, T.; Ramsey, G.; et al. Definition of regions in human c-Myc that are involved in transformation and nuclear localization. *Mol. Cell. Biol.* **1987**, *7*, 1697–1709.

207. Barrett, J.; Birrer, M. J.; Kato, G. J.; Dosaka-Akita, H.; Dang, C. V. Activation domains of L-Myc and c-Myc determine their transforming potencies in rat embryo cells. *Mol. Cell. Biol.* **1992**, *12*, 3130–3137.

208. Gilladoga, A. D.; Edelhoff, S.; Blackwood, E. M.; Eisenman, R. N.; Disteche, C. M. Mapping of Max to human chromosome 14 and mouse chromosome 12 by *in situ* hybridization. *Oncogene* **1992**, *7*, 1249–1251.

209. Wagner, A. J.; Le Beau, M. M.; Diaz, M. O.; Hay, N. Expression, regulation, and chromosomal localization of the *max* gene. *Proc. Natl. Acad. Sci. USA* **1992**, *89*, 3111–3115.

210. Begley, C. G.; Aplan, P. D.; Denning, S. M.; Haynes, B. F.; Waldmann, T. A.; Kirsch, I. R. The gene *scl* is expressed during early hematopoiesis and encodes a differentiation-related DNA-binding motif. *Proc. Natl. Acad. Sci. USA* **1989**, *86*, 10128–10132.

211. Green, A. R.; Salvaris, E.; Begley, C. G. Erythroid expression of the "helix-loop-helix" gene *scl*. *Oncogene* **1991**, *6*, 475–479.

212. Chen, Q.; Cheng, J.-T.; Tsai, L.-H.; et al. The *tal* gene undergoes chromosome translocation in T-cell leukemia and potentially encodes a helix-loop-helix protein. *EMBO J.* **1990**, *9*, 415–424.

213. Brown, L.; Cheng, J.-T.; Chen, Q.; et al. Site-specific recombination of the *tal-1* gene is a common occurrence in human T-cell leukemia. *EMBO J.* **1990**, *9*, 3343–3351.

214. Finger, L. R.; Kagan, J.; Christopher, G.; et al. Involvement of the *tcl5* gene on human chromosome 1 in T-cell leukemia and melanoma. *Proc. Natl. Acad. Sci. USA* **1989**, *86*, 5039–5043.

215. Mellentin, J. D.; Smith, S. D.; Cleary, M. D. *lyl-1*, a novel gene altered by chromosomal translocation in T cell acute leukemia, codes for a protein with a helix-loop-helix DNA binding motif. *Cell* **1989**, *58*, 77–83.

216. Call, K. M.; Glaser, T.; Ito, C. Y.; et al. Isolation and characterization of a zinc-finger polypeptide gene at the human chromosome 11 Wilm's tumor locus. *Cell* **1990**, *60*, 509–520.

217. Gessler, M.; Poustka, A.; Cavenee, W.; Neve, R. L.; Orkin, R. L.; Bruns, G. A. P. Homozygous deletion in Wilm's tumors of a zinc-finger gene identified by chromosome jumping. *Nature* **1990**, *343*, 774–778.

218. Mermod, N.; O'Neill, E. A.; Kelly, R.; Tjian, R. The proline-rich transcriptional activator of CTF/NFI is distinct from the replication and DNA binding domain. *Cell* **1989**, *58*, 741–753.

219. Tanaka, M.; Herr, W. Differential transcriptional activation by *oct-1* and *oct-2*: interdependent activation domains induce *oct-2* phosphorylation. *Cell* **1990**, *60*, 375–386.

220. Rauscher, F. J. III.; Morris, J. F.; Tournay, O. E.; Cook, D. M.; Curran, T. Binding of the Wilms' tumor locus zinc finger protein to the EGR-1 consensus sequence. *Science* **1990**, *250*, 1259–1261.

221. Lemaire, P.; Vesque, C.; Schmitt, J.; Stunnenberg, H.; Frank, R.; Charnay, P. The serum-inducible mouse gene *krox-24* encodes a sequence-specific transcriptional activator. *Mol. Cell. Biol.* **1990**, *10*, 3456–3467.

222. Patwardhan, S.; Gashler, A.; Siegel, M. G.; et al. EGR3, a novel member of the *egr* family of genes encoding immediate-early transcription factors. *Oncogene* **1991**, *6*, 917–928.

223. Madden, S. L.; Cook, D. M.; Morris, J. F.; Gashler, A.; Sukhatme, V. P.; Rauscher, F. J. III. Transcriptional repression mediated by the *wt1* Wilms tumor gene product. *Science* **1991**, *253*, 1550–1553.

224. Pritchard-Jones, K.; Fleming, S.; Davidson, D.; et al. The candidate Wilms' tumor gene is involved in genitourinary development. *Nature* **1990**, *346*, 194–197.

225. Pelletier, J.; Schalling, M.; Buckler, A. J.; Rogers, A.; Haber, D.A.; Housman, D. Expression of the Wilms' tumor gene *wt1* in the murine urogenital system. *Genes Dev.* **1991**, *5*, 1345–1356.

226. Weissman, B. E.; Saxon, P. J.; Pasquale, S. R.; Jones, G. R.; Geiser, A. G.; Stanbridge, E. J. Introduction of a normal human chromosome 11 into a Wilms' tumor cell line controls its tumorigenic expression. *Science* **1987**, *236*, 175–180.

227. Drummond, I. A.; Madden, S. L.; Rohwer-Nutter, P.; Bell, G. I.; Sukhatme, V. P.; Rauscher, F. J. III. Repression of the insulin-like growth factor II gene by the Wilms' tumor suppressor WT1. *Science* **1992**, *257*, 674–677.

228. Kinzler, K. W.; Bigner, S. H.; Bigner, D. D.; et al. Identification of an amplified, highly expressed gene in a human glioma. *Science* **1987**, *236*, 70–73.

229. Kinzler, K. W.; Ruppert, J. M.; Bigner, S. H.; Vogelstein, B. The *gli* gene is a member of the Krüppel family of zinc finger proteins. *Nature (London)* **1988**, *332*, 371–374.

230. Schuh, R.; Aicher, W.; Gaul, U.; et al. A conserved family of nuclear proteins containing structural elements of the finger protein encoded by Krüppel, a *Drosophila* segmentation gene. *Cell* **1986**, *47*, 1025–1032.

231. Chowdhury, K.; Deutsch, U.; Gruss, P. A multigene family encoding several "finger" structures is present and differentially active in mammalian genomes. *Cell* **1987**, *48*, 771–778.

232. Vortkamp, A.; Gessler, M.; Grzeschik, K.-H. *GLI3* zinc finger gene interrupted by translocations in Greig syndrome families. *Nature (London)* **1991**, *352*, 539–540.

233. Mucenski, M. L.; Taylor, B. A.; Ihle, J. N.; et al. Identification of a common ecotropic viral integration site, Evi-1, in the DNA of AKXD murine myeloid tumors. *Mol. Cell. Biol.* **1988**, *8*, 301–308.

234. Morishita, K.; Parker, D. S.; Mucenski, M. L.; Jenkins, N. A.; Copeland, N. G.; Ihle, J. N. Retroviral activation of a novel gene encoding a zinc finger protein in IL-3-dependent myeloid leukemia cell lines. *Cell* **1988**, *54*, 831–840.

235. Bartholomew, C.; Morishita, K.; Askew, D.; et al. Retroviral insertions in the CB-1/Fim-3 common site of integration activate expression of the *Evi-1* gene. *Oncogene* **1989**, *4*, 529–534.

236. Perkins, A. S.; Fishel, R.; Jenkins, N. A.; Copeland, N. G. Evi-1, a murine zinc finger protooncogene, encodes a sequence-specific DNA-binding protein. *Mol. Cell. Biol.* **1991**, *11*, 2665–2674.

237. Perkins, A. S.; Mercer, J. A.; Jenkins, N. A.; Copeland, N. G. Patterns of *Evi-1* expression in embryonic and adult tissues suggest that *Evi-1* plays an important regulatory role in mouse development. *Development* **1991**, *111*, 479–487.

238. Sap, J.; Muñoz, A.; Damm, K.; et al. The c-ErbA protein is a high-affinity receptor for thyroid hormone. *Nature (London)* **1986**, *324*, 635–640.

239. Weinberger, C.; Thompson, C. C.; Ong, E. S.; Lebo, R.; Gruol, D. I.; Evans, R. M. The c-*ErbA* gene encodes a thyroid hormone receptor. *Nature (London)* **1986**, *324*, 641–646.

240. Evans, R. M. The steroid and thyroid hormone receptor superfamily. *Science* **1988**, *240*, 889–895.

241. Thompson, C. C.; Evans, R. M. Trans-activation by thyroid hormone receptors: Functional parallels with steroid hormone receptors. *Proc. Natl. Acad. Sci. USA* **1989**, *86*, 3494–3498.

242. Damm, K.; Thompson, C. C.; Evans, R. M. Protein encoded by v-*ErbA* functions as a thyroid hormone receptor antagonist. *Nature (London)* **1989**, *339*, 593–597.

243. Zenke, M.; Kahn, P.; Disela, C.; et al. v-erbA specifically suppresses transcription of the avian erythrocyte anion transporter (band 3) gene. *Cell* **1988**, *52*, 107–119.

244. Sheer, D.; Lister, T. A.; Amess, J.; Solomon, E. Incidence of the 15q+;17q- chromosome translocation in acute promyelocytic leukemia (APL). *Br. J. Cancer* **1985**, *52*, 55–58.

245. Borrow, J.; Goddard, A. D.; Sheer, D.; Solomon, E. Molecular analysis of acute promyelocytic leukemia breakpoint cluster region on chromosome 17. *Science* **1990**, *249*, 1577–1580.

246. Kakizuka, A.; Miller, W. H. Jr.; Umesono, K.; et al. Chromosomal translocation t(15:17) in human acute promyelocytic leukemia fuses RAR alpha with a novel putative transcription factor, PML. *Cell* **1991**, *66*, 663–674.

247. Freemont, P. S.; Hanson, I. M.; Trowsdale, J. A novel cysteine-rich sequence motif.(Lett) *Cell* **1991**, *64*, 484–484.

248. Brunk, B. P.; Martin, E. C.; Adler, P. N. *Drosophila* genes Posterior, sex combs and Suppressor two of zeste encode proteins with homology to the murine *bmi-1* oncogene. *Nature (London)* **1991**, *353*, 351–353.

249. van Lohuizen, M.; Frasch, M.; Wientjens, E.; Berns, A. Sequence similarity between the mammalian *bmi-1* protooncogene and the *Drosophila* regulatory genes *Psc* and *Su(z)2*. *Nature (London)* **1991**, *353*, 353–355.

250. van Lohuizen, M.; Verbeek, S.; Scheijen, B.; Wientjens, E.; van der Gulden, H.; Berns, A. Identification of cooperating oncogenes in Eμ-*myc* transgenic mice by provirus tagging. *Cell* **1991**, *65*, 737–752.

251. Haupt, Y.; Alexander, W. S.; Barri, G.; Klinken, S. P.; Adams, J. M. Novel zinc finger gene implicated as myc collaborator by retrovirally accelerated lymphomagenesis in *Eμ-myc* transgenic mice. *Cell* **1991**, *65*, 753–763.

252. Tagawa, M.; Sakamoto, T.; Shigemoto, K.; et al. Expression of novel DNA-binding protein with zinc finger structure in various tumor cells. *J. Biol. Chem.* **1990**, *265*, 20021–20026.

253. Goebl, M. G. The *bmi-1* and *mel-18* gene products define a new family of DNA-binding proteins involved in cell proliferation and tumorigenesis. (Lett) *Cell* **1991**, *66*, 623.

254. Boehm, T.; Baer, R.; Lavenir, I.; et al. The mechanism of chromosomal translocation t(11:14) involving the T-cell receptor Cδ locus on human chromosome 14q11 and a transcribed region of chromosome 11p15. *EMBO J.* **1988**, *7*, 385–394.

255. Boehm, T.; Foroni, L.; Kaneko, Y.; Perutz, M. F.; Rabbitts, T. H. The rhombotin family of cysteine-rich LIM-domain oncogenes: distinct members are involved in T-cell translocations to human chromosomes 11p15 and 11p13. *Proc. Natl. Acad. Sci. USA* **1991**, *88*, 4367–4371.

256. McGuire, E. A.; Hockett, R. D.; Pollock, K. M.; Bartholdi, M. F.; O'Brien, S. J.; Korsmeyer, S. J. The t(11:14)(p15;q11) in a T-cell acute lymphoblastic leukemia cell line activates multiple transcripts, including *ttg-1*, a gene encoding a potential zinc finger protein. *Mol. Cell. Biol.* **1989**, *9*, 2124–2132.

257. Greenberg, J. M.; Boehm, T.; Sofroniew, M. V.; et al. Segmental and developmental regulation of a presumptive T-cell oncogene in the central nervous system. *Nature (London)* **1990**, *344*, 158–160.

258. Scheidereit, C.; Cromlish, J. A.; Gerster, T.; et al. A human lymphoid-specific transcription factor that activates immunoglobulin genes is a homeobox protein. *Nature (London)* **1988**, *336*, 551–557.

259. Johnson, W. A.; Hirsh, J. Binding of a *Drosophila* POU-domain protein to a sequence element regulating gene expression in specific dopaminergic neurons. *Nature (London)* **1990**, *343*, 467–470.

260. Williams, D. L.; Look, A. T.; Melvin, S. L.; et al. New chromosomal translocations correlate with specific immunophenotypes of childhood acute lymphoblastic leukemia. *Cell* **1984**, *36*, 101–109.

261. Nourse, J.; Mellentin, J. D.; Galili, N.; et al. Chromosomal translocation t(1:19) results in synthesis of a homeobox fusion mRNA that codes for a potential chimeric transcription factor. *Cell* **1990**, *60*, 535–545.

262. Kamps, M. P.; Murre, C.; Sun, X.-H.; Baltimore, D. A new homeobox gene contributes the DNA binding domain of the t(1:19) translocation protein in preB-ALL. *Cell* **1990**, *60*, 547–555.

263. Moss, L. G.; Moss, J. B.; Rutter, W. J. Systematic binding analysis of the insulin gene transcriptional control region: insulin and immunoglobulin enhancers utilize similar transactivators. *Mol. Cell. Biol.* **1988**, *8*, 2620–2627.

264. Blatt, C.; Aberdam, D.; Schwartz, R.; Sachs, L. DNA rearrangement of a homeobox gene in myeloid leukemic cells. *EMBO J.* **1988**, *7*, 4283–4290.

265. Aberdam, D.; Negreanu, V.; Sachs, L.; Blatt, C. The oncogenic potential of an activated *hox-2.4* homeobox gene in mouse fibroblasts. *Mol. Cell. Biol.* **1991**, *11*, 5554–5557.

266. Hatano, M.; Roberts, C. W.; Minden, M.; Crist, W. M.; Korsmeyer, S. J. Deregulation of a homeobox gene, *hox-11*, by the t(10;14) in T-cell leukemia. *Science* **1991**, *253*, 79–82.

267. Dubé, I. D.; Kamel-Reid, S.; Yuan, C. C.; et al. A novel homeobox gene lies at the chromosome 10 breakpoint in lymphoid neoplasias with chromosomal translocation t(10;14). *Blood* **1991**, *78*, 2996–3003.

268. Kennedy, M. A.; Gonzalez-Sarmiento, R.; Kees, U. R.; et al. HOX11, a homeobox-containing T-cell oncogene on human chromosome 10q24. *Proc. Natl. Acad. Sci. USA* **1991**, *88*, 8900–8904.

269. Stephens, R. M.; Rice, N. R.; Hiebsch, R. R.; Bose, H. R.; Gilden, R. V. Nucleotide sequence of v-*rel*: the oncogene of the reticuloendotheliosis virus. *Proc. Natl. Acad. Sci. USA* **1983**, *80*, 6229–6233.

270. Gilmore, T. D.; Temin, H. M. V-rel oncoproteins in the nucleus and in the cytoplasm transform chicken spleen cells. *J. Virol.* **1988**, *62*, 703–714.

271. Kieran, M.; Blank, V.; Logeat, F.; et al. The DNA-binding subunit of NF-κB is identical to factor KBF1 and homologous to the rel oncogene product. *Cell* **1990**, *62*, 1007–1018.

272. Ghosh, S.; Gifford, A. M.; Riviere, L. R.; Tempst, P.; Nolan, G. P.; Baltimore, D. Cloning of the p50 subunit of NF-κB: homology to *rel* and *dorsal*. *Cell* **1990**, *62*, 1019–1029.

273. Hannink, M.; Temin, H. M. Transactivation of gene expression by nuclear and cytoplasmic rel proteins. *Mol. Cell. Biol.* **1989**, *9*, 4323–4336.

274. Ballard, D. W.; Walker, W. H.; Doerre, S.; et al. The v-*rel* oncogene encodes a κB enhancer binding protein that inhibits NF-κB function. *Cell* **1990**, *63*, 803–814.

275. Narayanan, R.; Klement, J. F.; Ruben, S. M.; Higgins, K. A.; Rosen, C. A. Identification of a naturally occurring transforming variant of the p65 subunit of NF-κB. *Science* **1992**, *256*, 367–370.

276. Neri, A.; Chang, C. C.; Lombardi, L.; et al. B cell lymphoma-associated chromosomal translocation involves candidate oncogene *lyt-10*, homologous to NF-kappa B p50. *Cell* **1991**, *67*, 1075–1087.

277. Hatada, E. N.; Nieters, A.; Wulczyn, F. G.; et al. The ankyrin repeat domains of the NF-κB precursor p105 and the protooncogene *bcl-3* act as specific inhibitors of NF-κB DNA binding. *Proc. Natl. Acad. Sci. USA* **1992**, *89*, 2489–2493.

278. Gonda, T. J.; Bishop, J. M. Structure and transcription of the cellular homolog (c-*myb*) of the avian myeloblastosis virus transforming gene (v-myb). *J. Virol.* **1983**, *46*, 212–220.

279. Klempnauer, K.-H.; Ramsay, G.; Bishop, J. M.; et al. The product of the retroviral transforming gene, v-*myb*, is a truncated version of the protein encoded by the cellular oncogene c-*myb*. *Cell* **1983**, *33*, 345–355.

280. Klempnauer, K.-H.; Symonds, G.; Evan, G. I.; Bishop, J. M. Subcellular localization of proteins encoded by oncogenes of avian myeloblastosis virus and avian leukemia virus E26 and the chicken c-*myb* gene. *Cell* **1984**, *37*, 537–547.

281. Klempnauer, K.-H.; Bonifer, C.; Sippel, A. E. Identification and characterization of the protein encoded by the human c-*myb* protooncogene. *EMBO J.* **1986**, *5*, 1903–1911.

282. Duprey, S. P.; Boettiger, D. Developmental regulation of c-*myb* in normal myeloid progenitor cells. *Proc. Natl. Acad. Sci. USA* **1985**, *82*, 6937–6941.

283. Sakura, A.; Kanei-Ishii, C.; Nagase, T.; Nakagoshi, H.; Gonda, T.J.; Ishii, S. Delineation of three functional domains of the transcriptional activator encoded by the c-*myb* protooncogene. *Proc. Natl. Acad. Sci. USA* **1989**, *85*, 5758–5762.

284. Weston, K.; Bishop, J. M. Transcriptional activation by the v-myb oncogene and its cellular progenitor c-*myb*. *Cell* **1989**, *58*, 85.

285. Gazzolo, L.; Moscovici, C.; Moscovici, M. G. Response of haemopoietic cells to avian acute leukemia viruses: effects on the differentiation of the target cells. *Cell* **1979**, *16*, 627–638.

286. McMahon, J.; Howe, K. M.; Watson, R. J. The induction of Friend erythroleukemia differentiation is markedly affected by expression of a transfected c-*myb* cDNA. *Oncogene* **1988**, *3*, 717–720.

287. Howe, K. M.; Reakes, C. F. L.; Watson, R. J. Characterization of the sequence-specific interaction of mouse c-Myb protein with DNA. *EMBO J.* **1990**, *9*, 161–169.

288. Gabrielsen, O. S.; Sentenac, A.; Fromageot, P. Specific DNA binding by c-Myb: Evidence for a double helix-turn-helix-related motif. *Science* **1991**, *253*, 1140–1143.

289. Ogata, K.; Hojo, H.; Aimoto, S.; et al. Solution structure of a DNA-binding unit of Myb: A helix-turn-helix-related motif with conserved tryptophans forming a hydrophobic core. *Proc. Natl. Acad. Sci. USA* **1992**, *89*, 6428–6432.

290. Ness, S. A.; Marknell, A.; Graf, T. The v-myb oncogene product binds to and activates the promyelocyte-specific *mim-1* gene. *Cell* **1989**, *59*, 1115–1125.

291. Reiss, K.; Ferber, A.; Travali, S.; Porcu, P.; Phillips, P. D.; Baserga, R. The protooncogene c-*myb* increases the expression of insulin-like growth factor 1 and insulin-like growth factor 1 receptor messenger RNAs by a transcriptional mechanism. *Cancer Res.* **1991**, *51*, 5997–6000.

292. Leprince, D.; Gegonne, A.; Coll, C.; et al. A putative second cell-derived oncogene of the avian leukemia retrovirus E26. *Nature* **1983**, *306*, 395–397.

293. Nunn, M. F.; Seeburg, P. H.; Moscovici, C.; Duesburg, P. H. Tripartite structure of the avian erythroblastosis virus E26 transforming gene. *Nature (London)* **1986**, *306*, 391–395.

294. Nunn, M. F.; Hunter, T. The ets sequence is required for induction of erythroblastosis in chickens by avian retrovirus E26. *J. Virol.* **1989**, *63*, 398–402.

295. Fujiwara, S.; Fisher, R. J.; Seth, A.; et al. Characterization and localization of the products of the human homologs of the v-ets oncogene. *Oncogene* **1988**, *2*, 99–103.

296. Fujiwara, S.; Fisher, R. J.; Bhat, N. K.; Espina, S.; Papas, T. S. A short lived nuclear phosphoprotein encoded by the human *ets-2* protooncogene is stabilized by activation of protein kinase C. *Mol. Cell. Biol.* **1988**, *8*, 4700–4706.

297. Pognonec, P.; Boulukos, K. E.; Ghysdael, J. The c-ets-1 protein is chromatin associated and binds to DNA. *Oncogene* **1989**, *4*, 691–697.

298. Watson, D. K.; McWilliams, M. J.; Lapis, P.; Lautenberger, J. A.; Schweinfest, C. W.; Papas, T. S. Mammalian *ets-1* and *ets-2* genes encode highly conserved proteins. *Proc. Natl. Acad. Sci. USA* **1988**, *85*, 7862–7866.

299. Pribyl, L. J.; Watson, D. K.; McWilliams, M. J.; Ascione, R.; Papas, T. S. The *Drosophila ets-2* gene: Molecular structure, chromosomal localization, and developmental expression. *Dev. Biol.* **1988**, *127*, 45–53.

300. Reddy, E. S. P.; Rao, V. N.; Papas, T. S. The *erg* gene: A human gene related to the *ets* gene. *Proc. Natl. Acad. Sci. USA* **1987**, *84*, 6131–6135.

301. Rao, V. N.; Huebner, K.; Isobe, M.; Ar-Rushdi, A.; Croce, C. M.; Reddy, E. S. P. elk, tissue-specific *ets*-related genes on chromosomes X and 14 near translocation breakpoints. *Science* **1989**, *244*, 66–70.

302. Gunther, C. V.; Nye, J. A.; Bryner, R. S.; Graves, B. J. Sequence-specific DNA binding of the protooncogene *ets-1* defines a transcriptional activator sequence within the long terminal repeat of the Moloney murine sarcoma virus. *Genes Dev.* **1990**, *4*, 667–679.

303. Gitlin, S. D.; Bosselut, R.; Gegonne, A.; Ghysdael, J.; Brady, J. N. Sequence-specific interaction of the Ets1 protein with the long terminal repeat of the human T-lymphotropic virus type 1. *J. Virol.* **1991**, *65*, 5513–5523.

304. Mavrothalassitis, G. J.; Papas, T. S. Positive and negative factors regulate the transcription of the *ets-2* gene via an oncogene-responsive-like unit within the *ets-2* promoter region. *Cell Growth Differ.* **1991**, *2*, 215–224.

305. Ho, I. C.; Bhat, N. K.; Gottschalk, L. R.; et al. Sequence-specific binding of human *ets-1* to the T cell receptor alpha gene enhancer. *Science* **1990**, *250*, 814–818.

306. Wasylyk, B.; Wasylyk, C.; Flores, P.; Begue, A.; Leprince, D.; Stehelin, D. The c-*ets* protooncogenes encode transcription factors that cooperate with c-*fos* and c-*jun* for transcriptional activation. *Nature* **1990**, *346*, 191–193.

307. Wasylyk, C.; Flores, P.; Gutman, A.; Wasylyk, B. PEA3 is a nuclear target for transcription activation by non-nuclear oncogenes. *EMBO J.* **1989**, *8*, 3371–3378.

308. Bruder, J. T.; Heidecker, G.; Rapp, U. R. Serum-, TPA-, and Ras-induced expression from AP-1/*ets*-driven promoters requires *raf-1* kinase. *Genes Dev.* **1992**, *6*, 545–556.

309. Bhat, N. K.; Thompson, C. B.; Lindsten, T.; et al. Reciprocal expression of human *ets-1* and *ets-2* genes during T-cell activation: regulatory role for the protooncogene *ets-1*. *Proc. Natl. Acad. Sci. USA* **1990**, *87*, 3723–3727.

310. Wernert, N.; Raes, M. B.; Lassalle, P.; et al. c-ets1 protooncogene is a transcription factor expressed in endothelial cells during tumor vascularization and other forms of angiogenesis in humans. *Am. J. Pathol.* **1992**, *140*, 119–127.

311. Metz, T.; Graf, T. The nuclear oncogenes v-*erbA* and v-*ets* cooperate in the induction of avian erythroleukemia. *Oncogene* **1992**, *7*, 597–605.

312. Ben-David, Y.; Giddens, E. B.; Letwin, K.; Bernstein, A. Erythroleukemia induction by Friend murine leukemia virus: Insertional activation of a new member of the *ets* gene family, *fli-1*, closely linked to c-*ets-1*. *Genes Dev.* **1991**, *5*, 908–918.

313. Pongubala, J. M. R.; Nagulapalli, S.; Klemsz, M. J.; McKercher, S. R.; Maki, R. A.; Atchison, M. L. PU.1 recruits a second nuclear factor to a site important for immunoglobulin kappa 3′ enhancer activity. *Mol. Cell. Biol.* **1992**, *12*, 368–378.

314. Hipskind, R. A.; Rao, V. N.; Mueller, C. G.; Reddy, E. S.; Nordheim, A. Ets-related protein Elk-1 is homologous to the c-*fos* regulatory factor p62TCF. *Nature* **1991**, *354*, 531–534.

315. Dalton, S.; Treisman, R. Characterization of SAP-1, a protein recruited by serum response factor to the c-*fos* serum response element. *Cell* **1992**, *68*, 597–612.

316. Stavnezer, E.; Brodeur, D.; Brennan, L. The v-ski oncogene encodes a truncated set of c-*ski* coding exons with limited sequence and structural relatedness to v-*myc*. *Mol. Cell. Biol.* **1989**, *9*, 4038–4045.

317. Colmenares, C.; Stavnezer, E. The *ski* oncogene induces muscle differentiation in quail embryo cells. *Cell* **1989**, *59*, 293–303.

318. Sutrave, P.; Hughes, S. H. The ski oncogene. *Oncogene* **1991**, *6*, 353–356.

319. Friend, S. H.; Bernards, R.; Rogelj, S.; et al. A human DNA sequence with properties of the gene that predisposes to retinoblastoma and osteosarcoma. *Nature (London)* **1986**, *323*, 643–646.

320. Fung, Y.-K. T.; Murphree, A. L.; T'Ang, A.; Qian, J.; Benedict, W. F. Structural evidence for the authenticity of the human retinoblastoma gene. *Science* **1987**, *236*, 1657–1661.

321. Lee, W.-H.; Shew, J.-Y.; Hong, F. D.; et al. The retinoblastoma susceptibility gene encodes a nuclear phosphoprotein associated with DNA binding activity. *Nature* **1987**, *329*, 642–645.

322. Lee, E. Y.-H. P.; To, H.; Shew, J.-Y.; Bookstein, R.; Scully, P.; Lee, W.-H. Inactivation of the retinoblastoma susceptibility gene in human breast cancers. *Science* **1988**, *241*, 218–221.

323. Harbour, J. W.; Lai, S.-L.; Whang-Peng, J.; Gazdar, A. F.; Minna, J. D.; Kaye, F. J. Abnormalities in structure and expression of the human retinoblastoma gene in SCLC. *Science* **1988**, *241*, 353–357.

324. Varley, J. M.; Armour, J.; Swallow, J. E.; et al. The retinoblastoma gene is frequently altered leading to loss of expression in primary breast tumours. *Oncogene* **1989**, *4*, 725–729.

325. Venter, D. J.; Bevan, K. L.; Ludwig, R. L.; et al. Retinoblastoma gene deletions in human glioblastomas. *Oncogene* **1991**, *6*, 445–448.

326. Whyte, P.; Buchkovich, K. J.; Horowitz, J. M.; et al. Association between an oncogene and an anti-oncogene: The adenovirus E1A proteins bind to the retinoblastoma gene product. *Nature (London)* **1988**, *334*, 124–129.

327. DeCaprio, J. A.; Ludlow, J. W.; Figge, J.; et al. SV40 large tumor antigen forms a specific complex with the product of the retinoblastoma susceptibility gene. *Cell* **1988**, *54*, 275–283.

328. Dyson, N.; Howley, P. M.; Munger, K.; Harlow, E. The human papillomavirus-16 E7 oncoprotein is able to bind to the retinoblastoma gene product. *Science* **1989**, *243*, 934–937.

329. Buchkovich, K.; Duffy, L. A.; Harlow, E. The retinoblastoma protein is phosphorylated during specific phases of the cell cycle. *Cell* **1989**, *58*, 1097–1105.

330. Chen, P. L.; Scully, P.; Shew, J.-Y.; Wang, J.-Y.; Lee, W.-H. Phosphorylation of the retinoblastoma gene product is modulated during the cell cycle and cellular differentiation. *Cell* **1989**, *58*, 1193–1198.

331. Kaelin, W. G. Jr.; Pallas, D. C.; DeCaprio, J. A.; Kaye, F. J.; Livingston, D. M. Identification of cellular proteins that can interact specifically with the T/E1A-binding region of the retinoblastoma gene product. *Cell* **1991**, *64*, 521–532.

332. Chellappan, S. P.; Hiebert, S.; Mudryj, M.; Horowitz, J. M.; Nevins, J. R. The E2F transcription factor is a cellular target for the RB protein. *Cell* **1991**, *65*, 1053–1061.

333. Chittenden, T.; Livingston, D. M.; Kaelin, W. G.Jr. The T/E1A-binding domain of the retinoblastoma product can interact selectively with a sequence-specific DNA-binding protein. *Cell* **1991**, *65*, 1073–1082.

334. Mudryj, M.; Hiebert, S. W.; Nevins, J. R. A role for the adenovirus inducible E2F transcription factor in a proliferation dependent signal transduction pathway. *EMBO J.* **1990**, *9*, 2179–2184.

335. Mudryj, M.; Devoto, S. H.; Hiebert, S. W.; Hunter, T.; Pines, J.; Nevins, J. R. Cell cycle regulation of the E2F transcription factor involves an interaction with cyclin A. *Cell* **1991**, *65*, 1243–1253.

336. Chellappan, S.; Kraus, V. B.; Kroger, B.; et al. Adenovirus E1A, simian virus 40 tumor antigen, and the human papillomavirus E7 protein share the capacity to disrupt the interaction between transcription factor E2F and the retinoblastoma gene product. *Proc. Natl. Acad. Sci. USA* **1992**, *89*, 4549–4553.

337. Robbins, P. D.; Horowitz, J. M.; Mulligan, R. C. Negative regulation of human c-*fos* expression by the retinoblastoma gene product. *Nature (London)* **1990**, *346*, 668–671.

338. Rustgi, A. K.; Dyson, N.; Bernards, R. Amino-terminal domains of c-myc and N-myc proteins mediate binding to the retinoblastoma gene product. *Nature (London)* **1991**, *352*, 541–544.

339. Kim, S.-J.; Wagner, S.; Liu, F.; O'Reilly, M. A.; Robbins, P. D.; Green, M. R. Retinoblastoma gene product activates expression of the human *TGF*–β2 gene through transcription factor ATF-2. *Nature (London)* **1992**, *358*, 331–334.

340. Lane, D. P.; Gannon, J. Cellular proteins involved in SV40 transformation. *Cell. Biol. Int. Rep.* **1983**, *7*, 513–514.

341. Sarnow, P.; Ho, Y.-S.; Williams, J.; Levine, A. J. Adenovirus E1b-58kd tumor antigen and SV40 tumor antigen are physically associated with the same 54-kDa cellular protein in transformed cells. *Cell* **1982**, *26*, 387–394.

342. Lane, D. P.; Crawford, L. V. T antigen is bound to host protein in SV40-transformed cells. *Nature (London)* **1979**, *278*, 261–263.

343. Linzer, D. I. H.; Levine, A. J. Characterization of a 54Kda cellular SV40 tumor antigen present in SV40-transformed cells and uninfected embryonal cells. *Cell* **1979**, *17*, 43–52.

344. Werness, B. A.; Levine, A. J.; Howley, P. M. Association of human papillomavirus types 16 and 18 E6 proteins with p53. *Science* **1990**, *248*, 76–79.

345. Baker, S. J.; Markowitz, S.; Fearon, E. R.; Willson, J. K. V.; Vogelstein, B. Suppression of human colorectal carcinoma cell growth by wild-type p53. *Science* **1990**, *249*, 912–915.

346. Diller, L.; Kassel, J.; Nelson, M. A.; et al. p53 functions as a cell cycle control protein in osteosarcomas. *Mol. Cell. Biol.* **1990**, *10*, 5775–5781.

347. Mercer, W. E.; Shields, M. T.; Amin, M.; et al. Negative growth regulation in a glioblastoma tumor cell line that conditionally expresses human wild-type p53. *Proc. Natl. Acad. Sci. USA* **1990**, *87*, 6166–6170.

348. Hinds, P. W.; Finlay, C. A.; Quartain, R. S.; et al. Mutant p53 cDNAs from human colorectal carcinomas can cooperate with Ras in transformation of primary rat cells. *Cell Growth Differ.* **1990**, *1*, 571–580.

349. Pennica, D.; Goeddel, D. V.; Hayflick, J. S.; Reich, N. C.; Anderson, C. W.; Levine, A. J. The amino acid sequence of murine p53 determined from a cDNA clone. *Virology* **1984**, *134*, 477–482.

350. Fields, S.; Jang, S. K. Presence of a potent transcription activating sequence in the p53 protein. *Science* **1990**, *249*, 1046–1049.

351. Raycroft, L.; Wu, H.; Lozano, G. Transcriptional activation by wild-type but not transforming mutants of the *p53* anti-oncogene. *Science* **1990**, *249*, 1049–1051.

352. Raycroft, L.; Schmidt, J. R.; Yoas, K.; Hao, M.; Lozano, G. Analysis of p53 mutants for transcriptional activity. *Mol. Cell. Biol.* **1991**, *11*, 6067–6074.

353. Santhanam, U.; Ray, A.; Sehgal, P. B. Repression of the interleukin 6 gene promoter by the *p53* and retinoblastoma susceptibility gene product. *Proc. Natl. Acad. Sci. USA* **1991**, *88*, 7605–7609.

354. Ginsberg, D.; Mechta, F.; Yaniv, M.; Oren, M. Wild-type p53 can down-modulate the activity of various promoters. *Proc. Natl. Acad. Sci. USA* **1990**, *88*, 9979–9983.

355. Mercer, W. E.; Shields, M. T.; Lin, D.; Appella, E.; Ullrich, S. J. Growth suppression induced by wild-type p53 protein is accompanied by selective down-regulation of proliferating-cell nuclear antigen expression. *Proc. Natl. Acad. Sci. USA* **1991**, *88*, 1958–1962.
356. Bargonetti, J.; Friedman, P. N.; Kern, S. E.; Vogelstein, B.; Prives, C. Wild-type but not mutant p53 immunopurified proteins bind to sequences adjacent to the SV40 origin of replication. *Cell* **1991**, *65*, 1083–1091.
357. Kern, S. E.; Kinzler, K. W.; Bruskin, A.; et al. Identification of p53 as a sequence specific DNA-binding protein. *Science* **1991**, *252*, 1708–1711.
358. Zambetti, G. P.; Bargonetti, J.; Walker, K.; Prives, C.; Levine, A. J. Wild-type p53 mediates positive regulation of gene expression through a specific DNA sequence element. *Genes Dev.* **1992**, *6*, 1143–1152.
359. Kern, S. E.; Pietenpol, J. A.; Thiagalingam, S.; Seymour, A.; Kinzler, K. W.; Vogelstein, B. Oncogenic forms of p53 inhibit *p53*-regulated gene expression. *Science* **1992**, *256*, 827–830.
360. Kafiri, G.; Thomas, D. M.; Shepherd, N. A.; Krausz, T.; Lane, D. P.; Hall, P. A. *p53* expression is common in malignant mesothelioma. *Histopathol.* **1992**, *21*, 331–334.
361. Sawan, A.; Randall, B.; Angus, B.; et al. Retinoblastoma and *p53* gene expression relate to relapse and survival in human breast cancer: An immunohistochemical study. *J. Pathol.* **1992**, *168*, 23–28.

LOSS OF CONSTITUTIONAL HETEROZYGOSITY IN HUMAN CANCER:

A PRACTICAL APPROACH

Jan Zedenius, Günther Weber, and Catharina Larsson

Advances in Genome Biology
Volume 3B, pages 279–303.
Copyright © 1995 by JAI Press Inc.
All rights of reproduction in any form reserved.
ISBN: 1-55938-835-8

I. INTRODUCTION

Human cancer arises from irreversible alterations within the genetic content of a single cell that are inherited by its daughter cells. These mutations may activate the so-called proto-oncogenes, causing their overexpression or hyperfunction, and thereby lead to uncontrolled cell growth. Additionally, it has now become evident that, apart from this activation of oncogenes, loss of genetic information may contribute to cancer development. First observed in rare heritable cancer syndromes and eliminating genes that exert cellular control functions, "genetic losses" have now also been found in frequent and non-inherited cancers as the initial step of tumorigenesis and contributing to tumor progression.

Screening for genetic losses is therefore a generally applied method for the characterization of cancer. Apart from methods related to classical cytogenetics, new molecular tools developed in the past decade have dramatically refined the resolution and facilitated the characterization of the tumor cell genotype. DNA markers have been developed that distinguish between the alleles of homologous chromosomes, enabling the investigator to monitor losses of single alleles, and to determine their parental origin. Thus, detection of "loss of heterozygosity" (LOH) has become a widespread method in cancer research, and in the search for the specific genes involved in the genesis and progression of several tumors.

In the following, we will give the background to the usage of the LOH technique, try to describe some major technical and interpretative pitfalls when using it, and to give examples of genetic diseases where LOH has decisively contributed to the description of the mechanisms behind them.

II. CANCER: A GENETIC DISEASE

The hypothesis that cancer is a malady of genes is based on the following observations:

- *Familial aggregation of specific types of tumors.* This has been well documented, and familial clustering has been reported in more than 200 different neoplasms or syndromes.[1,2]
- *Constitutional chromosomal aberrations that confer an increased risk of developing specific tumors.* For example, patients suffering from the rare eye tumor in children, retinoblastoma (RB), sometimes show a constitutional deletion of chromosomal region 13q14.[3,4]

- *Chromosomal rearrangements in tumor tissue.* This was first suggested by Boveri in 1914.[5] Development of the QFQ-banding technique in 1970 permitted detailed karyotype analysis.[6] Specific chromosomal aberrations could then be regularly associated with certain types of neoplasias, the classical example being the Philadelphia chromosome translocation in chronic myeloid leukemia.[7]
- *Carcinogenic agents are mutagenic.* Different tumor forms have been associated with environmental factors, such as smoking, sunlight, and dietary components. Since most carcinogenic agents also are known to be mutagenic in experimental systems, it is in agreement with the idea that cancer can arise from somatic mutations.[8]
- *Constitutional deficiencies in the systems for repair of DNA damage predispose to tumor development.*[9] Patients with xeroderma pigmentosum are deficient in their repair of UV-induced DNA damage. They frequently develop skin cancers in sunlight exposed areas of the body, and their cells demonstrate increases of UV-induced mutations *in vitro.*
- *Identification of two classes of genes involved in tumor development.* The oncogenes need to be somatically activated to contribute to tumor growth, while the tumor suppressor genes exert their phenotypic effect by their functional inactivation.

III. ONCOGENES

Initially, oncogenes were identified as transforming components of retroviruses.[10,11] Human nontransforming counterparts were then discovered, the protooncogenes, generally believed to be involved in cell proliferation and differentiation. As a result of specific genetic alterations they may gain the capacity to transform a cell into a neoplastic state. These genes act dominantly on the cellular level since only one of the two gene copies has to be activated to transform a cell. The effects are mediated either through overexpression of the normal, or expression of an aberrant protein.[10,11] These phenomena may be caused by gene amplification, translocation and/or deletion of chromosomal regions, or point mutations within the actual gene. The oncogenes are classified according to the subcellular localization and biochemical function of their products. Some are, for example, found in the nucleus recognized as transcription factors, while others are found on the outside of the cell surface (i.e., growth factors bound to specific membrane receptors).

IV. TUMOR SUPPRESSOR GENES

In 1914, Boveri first suggested the existence of "inhibiting chromosomes," with the function to inhibit tumorigenesis.[5] The idea that certain familial cancer syn-

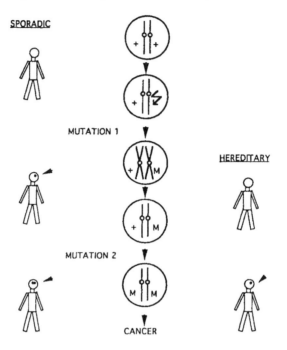

Figure 1. Model of the recessive mechanism of tumorigenesis. In patients with heritable cancer the first mutation is present in all cells. These are therefore predisposed to tumor development, but appear normal. In order for the tumor to develop, a second mutation which eliminates the normal allele/function (+) in one cell, has to occur. Even if the mutation frequency is low, the number of cells is so large that such a mutation most probably will occur in at least one cell. Hence, predisposed individuals typically develop multiple tumors with comparatively early age of onset. However, in sporadic cases, the two mutations must occur in the same somatic cell. This is less likely, and explains why sporadic tumors usually are single and occur later.

dromes result from additional somatic mutations in the individual who already carries a germline mutation was proposed by de Mars in 1970.[12] Shortly thereafter, Knudson's epidemiological analyses of RB led him to propose that development of the condition requires two independent mutations.[13] This theory was later generalized,[14] and is usually referred to as the "two-mutation model" for tumorigenesis (Figure 1).

The first evidence for the existence of tumor suppressor genes came from studies of somatic cell hybrids. In 1969, Harris showed that malignancy could be suppressed by the fusion of malignant and nonmalignant cells.[15] This inspired several research groups to search for tumor suppressor genes. The first and probably the best studied example is the RB gene (*rb*). The cloning of this gene permitted the experiment that provided direct evidence for the existence of a tumor suppressor gene, as the introduction of a cloned wild-type *rb* gene via a retroviral vector into RB cells resulted in suppression of the malignant phenotype.[16]

Several other tumor suppressor genes have since been localized and cloned. Examples are the recently cloned genes predisposing to familial adenomatous polyposis (*apc* gene),[17] neurofibromatosis type 1,[18–20] and neurofibromatosis type 2,[21] which were first mapped by linkage analysis in families segregating the disease.[22–26] The Wilms' tumor gene[27,28] was mapped by identification of constitutional chromosomal deletions.[29] Studies of genetic losses in colon cancers led to the cloning of the *dcc* gene, and also identified the *p53* gene to be involved in tumor progression.[30,31]

The *p53* story has truly been puzzling: it was first considered to be a tumor antigen, then a nuclear oncogene, and presently considered as a tumor suppressor gene.[32–34] In the response of a normal cell to DNA damage, the genome guarding function of the wild-type *p53* is induced.[35] This leads to cell cycle arrest in G1, enabling DNA repair or inducing apoptosis (cell death).[36] Disturbances in this function may lead to tumor growth. It is now clarified that the *p53* gene can also be involved in tumor initiation. In families with the rare Li–Fraumeni syndrome, associated with an increased risk for different cancers, the disease was found to segregate with a point mutation within the *p53* gene.[37,38]

In all examples of suppressor genes mentioned here, as well in the mapping of others, such as MEN1, meningioma, and breast carcinoma, a useful tool has been identification of chromosomal alterations by studying the LOH phenomenon in tumor tissue.[39–44]

V. PRINCIPLES OF LOSS OF HETEROZYGOSITY

The gateway to the studies of LOH was the chromosomal rearrangements seen in cytogenetic studies (i.e., karyotyping) of tumors. Karyotyping makes it possible to detect several types of aberrations, such as numerical deviations, translocations, inversions, large deletions, homogeneously staining regions, and double minutes (Figure 2). However, the method has several drawbacks. Since the cells must be studied in metaphase, the method demands growing cells. Karyotyping of solid tumors often involves tissue culturing, which may lead to selection of cells with a growth advantage *in vitro*, or the introduction of secondary aberrations that were not present in the primary tumor. Furthermore, the level of resolution is rather limited.

A general molecular genetic approach to study chromosomal rearrangements in tumors was introduced in 1983 by Cavenee and co-workers.[45] Here, by comparing constitutional and tumor genotypes, loss of genetic material in retinoblastomas were identified. This method has since been widely applied to several other tumor forms. Comprehensive summaries of allele losses frequently found in different

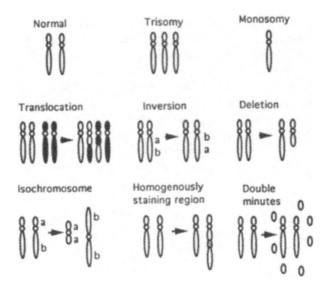

Figure 2. Examples of chromosome abnormalities, demonstrable by karyotyping.

tumors are available.[46,47] In the following we discuss how to detect allele losses, what significance they have for tumor development, and the genetic events they may reflect.

A. DNA Polymorphisms

Sequence variations in the human DNA occur every 300 to 500 base pairs (bp), within genes or between them.[48,49] There are two main groups of DNA polymorphisms: single bp alterations and variations in the number of repeat elements.

In approximately 5% of the cases, single bp alterations involve the cleavage site of a restriction endonuclease. These enzymes have bacterial origin, and recognize specific bp combinations, usually four or six bases, where they cleave the DNA. Polymorphic restriction sites result in restriction fragment length polymorphism (RFLP), i.e., different lengths of the restriction fragments. These fragments can be considered as alleles for the RFLP locus, and segregate in a codominant manner in families. In case of heterozygosity, one of the fragments represent the maternally, and the other the paternally derived chromosome.

The other main group of DNA polymorphisms are due to variations of the number of tandem repeats: the variable number of tandem repeats (VNTRs) usually reflect variations in repeat motifs of 10 to 20 bases per unit,[50,51] while the so called micro-satellite markers reflect sequence variations of shorter repeats, usually consisting of two to four nucleotide motifs, e.g., "CA-repeats".[52,53] The VNTRs and short microsatellite repeats are scattered over the entire human genome. Each array of repeats is flanked by unique DNA sequences. These unique (single copy) loci located close to a highly polymorphic repeat, constitute a set of very useful DNA markers.

Detection of RFLPs and VNTRs demands Southern blot analysis.[54] In short, the cleaved DNA fragments are size-separated by agarose gel electrophoresis. After denaturation, which makes the DNA single stranded, the fragments are transferred to a DNA binding filter. The filter is then used for hybridization with radio labeled fragments of genomic or complementary DNA probes. The hybridizing fragments are subsequently detected by autoradiography.

Detection of micro-satellite repeats requires very small amounts of DNA, since the alleles are amplified by a polymerase chain reaction (PCR), and then separated by polyacrylamide gel electrophoresis. PCR is a cyclic reaction used for specific amplification of DNA fragments between two known sequences. Each cycle includes denaturation, annealing of primers, and elongation of the strands. The primers are known complementary sequences used to direct the DNA polymerase enzyme, and the reaction is repeated in 20–40 cycles. The method is particularly useful for amplification of short DNA fragments (<1 kb), DNA from paraffin embedded tissues, and when only small amounts of cells are available.

B. Genetic Alterations Detectable by LOH

LOH can only be detected when the DNA marker (probe) is informative, that is, when the alleles show two separate bands on the autoradiogram. RFLP analysis demands a proper choice of restriction enzyme for the DNA probe in question.[55] In general, micro-satellite markers have the highest information content (sometimes more than 10 distinguishable alleles), followed by VNTR probes.

In the initial retinoblastoma studies by Cavenee et al., cloned DNA segments homologous to arbitrary sequences along chromosome 13, and which revealed RFLPs, were used to distinguish the alleles representing the two parental chromosomes.[45] The constitutional and tumor genotypes were compared at several loci, and the results indicated some of the putative second mutations, eliminating the normal function of a tumor suppressor gene (Figure 3).

LOH on all informative loci along a chromosome is usually interpreted as loss of one chromosome complement (Figure 3a and b). From the intensity of the remaining band/allele, the number of copies of the remaining chromosome is determined. For example, a doubling of the remaining allele indicates that two copies of the chromosome have co-segregated during mitosis (Figure 3b). When LOH is detected at one or a few loci, in combination with others on the same chromosome showing retained heterozygosity, this indicates a deletion or mitotic recombination (Figure 3c and d). Again, this distinction is made by measuring the intensity of the remaining allele, i.e., by visual inspection and densitometry analy-

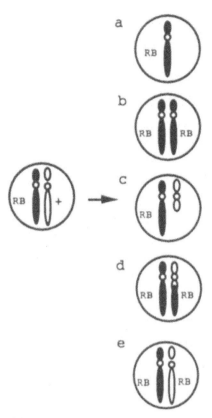

Figure 3. Examples of different putative second mutations eliminating the function of a tumor suppressor gene. (**a**) loss of a whole chromosome (**b**) imperfect cell division with co-segregation of the two copies of one of the homologs, or loss of one chromosome and duplication of the remaining (**c**) chromosome deletion (**d**) mitotic recombination (**e**) point mutation. Cytogenetic analysis would only detect **a** and possibly **c**, while molecular analysis can detect both **a**, **b**, **c**, **d**, and maybe **e** (if an intragenic marker is used).

sis. Point mutations are usually not detectable by LOH studies, unless they result in alteration of a restriction site (Figure 3e).

Some of these genetic alterations are demonstrable also by karyotyping, i.e., loss of a whole chromosome or large deletions. In contrast, chromosome loss coupled with duplication of the remaining chromosome, and also mitotic recombination, result in two apparently intact chromosomes, and can therefore not be detected by karyotyping. On the other hand, alterations such as translocations and inversions are readily detectable by karyotyping, while LOH is not feasible in this respect.

When using the technique for identification of LOH, one should be aware of additional genetic alterations that might occur. For example, *trisomy* may be identified on the autoradiogram as a doubled signal intensity for one of the alleles, but without loss of the other. In addition, *gene amplification* (e.g., of an oncogene) can be revealed as increased signal intensity for one or both alleles.[56]

C. Considerations when Performing LOH Studies

The experience in LOH studies so far obtained is mainly based upon Southern blot analysis. Basic requirements for this technique are the availability of fresh frozen tumors and corresponding constitutional DNA (e.g., from the patients leukocytes, fibroblasts, or normal tissue surrounding the tumor). The detection of micro-satellite repeats by CR makes it possible to analyze smaller amounts of DNA, and even DNA from paraffin embedded tissues. Today, more than 800 well-mapped, highly polymorphic CA-repeats evenly distributed over the human genome are available.[57] The disadvantages using the micro-satellite techniques are not yet well-defined. However, the comparatively long experience from LOH studies using Southern blotting gives us the opportunity to point out some pitfalls:

1.   Loss of genetic material must be distinguished from gain (Figure 4A, G). A frequent finding is that the alleles in the tumor lane differ in intensity relative to the constitutional alleles (i.e., allelic imbalance). This may indicate a true LOH, and/or doubling or amplification of the other allele. This is best distinguished if the amount of DNA loaded in each gel lane is even. From our own experience, we recommend that tumor DNA concentration judgement is based on both spectrophotometry as well as agarose gel electrophoresis evaluation. Standardization between the lanes can be done by comparison with a marker (a) with ascertained retain of heterozygosity, (b) detecting alleles of similar size, and (c) hybridized to the same filter.

2.   Correct exposure time of the autoradiogram is crucial. For instance, a short exposure time in combination with allelic imbalance and less DNA in the tumor lane, may give rise to false LOH. Similarly, overexposure may hide an existing allelic loss by saturation of the X-ray film (Figure 4F).

3.   Degraded tumor DNA may mimic LOH. This must be taken into account when there is a considerable size difference between the two alleles, and even more so if one of them is very large (Figure 4D).

4.   Partial restriction cleavage must be avoided, as this may also mimic LOH (Figure 4E). Note that restriction cleavage of DNA from solid tissue is less efficient than that in DNA from leukocytes. Furthermore, differences in methylation pattern must be taken into consideration, as the restriction enzymes differ in methylation sensitivity. Cleavage control should therefore be performed with parallel reactions including, for example, lambda DNA, in order to monitor specific phage band patterns on agarose gels.

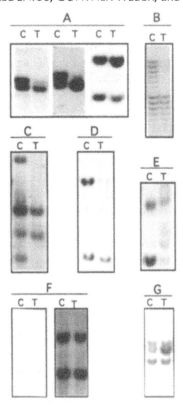

Figure 4. Examples of true LOH, and also some false positive and false negative results of LOH. (**A**) True LOH, detected in a sporadic breast carcinoma by RFLP analysis. (**B**) True LOH in a familial breast carcinoma, detected by PCR-amplification of a micro-satellite marker. (**C**) Vector contamination of constitutional DNA from a patient with a thyroid tumor. Where are the alleles? (**D**) False positive LOH due to degradation and partial cleavage of the tumor DNA. (**E**) False positive LOH due to partial cleavage of the tumor DNA. (**F**) False negative LOH due to overexposure of the autoradiogram. (**G**) Allelic imbalance reflecting trisomy 15 in a mouse plasmacytoma.

5. Plasmid contamination of DNA may also make reading of the results more difficult (Figure 4C). Where are the alleles? Are there constant bands? Are there bands from vector contamination? If the latter is suspected, it is possible to overcome the problem by isolating the insert of the probe before hybridization.

6. Contamination of normal tissue DNA in the tumor may cause a false negative result. Many tumor forms show a high grade of lymphocyte infiltration and/or large amounts of fibrous tissue. Others may show a growth pattern with sprouting of tumor cell formations into surrounding normal tissue. Therefore, it is suggested that a representative section is cut out from the tumor piece to be analyzed, and subjected

to histo-pathological examination. Tumor tissue containing less than 40–50% tumor cells is therefore not well-suited for studies of LOH. When using micro-satellite markers, a higher proportion of tumor cells is required. The amplification obtained by PCR is not linear, which may result in selection for normal tissue DNA.

7.   Clonality of the tumor cells must be considered. The alteration you are about to detect must be present in at least 40–50% of the cells studied, thus tumors of a polyclonal origin are less recommendable for LOH studies. For the same reason, subclonal events only present in a fraction of the tumor cells may escape detection by LOH analysis.

VI.   THE RETINOBLASTOMA PARADIGM

The first tumor suppressor gene identified was the *rb* gene responsible for a rare childhood tumor of the eye, retinoblastoma (RB). Ever since, RB has served as prototype for theories of the genetic mechanisms of cancer, and must be considered as the empirical proof for the two-mutation model proposed by Knudson.[13]

The first steps in the cloning of the *rb* gene was the localization of the disease locus to chromosomal region 13q14 by cytogenetic methods. A small percentage of RB patients were found to have constitutional deletions overlapping the region,[3,4] and tumor cells sometimes showed similar rearrangements of chromosome 13.[58] Family studies showed an association between the disease and an unbalanced translocation lacking the 13q14 region.[59] However, most RB families were cyto-genetically "normal," and in tumors the most frequent cytogenetic finding was isochromosome 6.

The close linkage of the disease gene to the genetic locus for esterase D (ESD) on 13q14, provided the possibility to study LOH on the protein level by distinguish-ing electrophoretic variants of ESD.[60] One study described a tumor that contained one chromosome 13 and had no ESD-activity, while the patient's constitutional level of ESD was only 50% of the normal. These findings suggested that the tumor had lost the normal chromosome 13 and provided the first experimental evidence that the *rb* gene is a recessive cancer gene.[61]

Cavenee's original studies of LOH in these tumors introduced a general approach to study the second event in tumorigenesis according to the two-mutation model. The constitutional and tumor genotypes were compared at several chromosome 13 loci, and the results indicated some of the possible types of second mutations suggested by Knudson (see Figure 3).[45,14]

LOH for chromosome 13 loci were found to occur in approximately half of all RBs. Allele losses were found in both sporadic and heritable forms of the disease, and were present in primary tumor tissue, cultured tumor cells, and tumors passaged through immunodeficient mice. They were also specific in the sense that other chromosomes tested were not found to be affected.[45,62]

RFLP markers were used to determine the parental origin of the rearranged or lost chromosome. As expected, the lost chromosome was always derived from the unaffected parent, while the one remaining carried the mutated retinoblastoma gene (i.e., originated from the affected parent).[63]

The key to the cloning of the *rb* gene was a probe (H3-8) which was derived from a flow sorted chromosome 13 specific library.[64–67] This marker was found to be homozygously deleted in two RB tumors, and a third tumor showed an interstitial deletion with one of the endpoints in vicinity of H3-8.[68] By locus expansion of H3-8, a conserved sequence was identified. Subsequently, corresponding cDNA clones were isolated and sequenced.[65,66] Gene deletions and absent or abnormal expression of the corresponding mRNA supported the authenticity of the gene.[65–67]

Finally, the experiment that provided direct evidence for the role of the *rb* gene in tumorigenesis could be performed. The introduction of a wild-type *rb* gene into tumor cells reverted the malignant phenotype.[16]

VII. THE MEN EXAMPLES

The two-mutational model for tumorigenesis implies that it would be possible to determine where the gene for the heritable form of a certain neoplasia is situated, simply by detecting genetic alterations which might reflect the second mutational event. Such information may be available from karyotypes of cultured tumor cells, or by studies of LOH.

LOH was used to localize the multiple endocrine neoplasia type 1 (MEN1) gene, an autosomal dominant predisposition to develop tumors in the parathyroids, the neuro-endocrine pancreas and duodenum, and the anterior pituitary.[69] Cytogenetic analysis of MEN1 patients had not revealed any constitutional chromosomal abnormalities that could aid to the localization of the MEN1 gene. However, if MEN1-associated tumors result from unmasking of a recessive mutation according to the two-mutation model, chromosomal rearrangements in such tumors might indicate which chromosome the MEN1 gene is situated on. Therefore, constitutional and tumor genotypes were compared at different RFLP loci in two brothers with neuro-endocrine pancreatic tumors who had inherited the disease from their mother. Markers on 17 chromosomes showed retained constitutional genotypes, but both tumors had lost one of the constitutional alleles at all informative loci on chromosome 11. The significance of these findings was further supported when the parental origin of the lost chromosome was determined (Figure 5).[39] In both cases, the lost alleles were always derived from the unaffected father. These findings did fit the hypothesis that the tumors resulted from elimination of the normal allele at the MEN1 locus, and that the MEN1 gene is a tumor suppressor gene located on chromosome 11. Subsequently, MEN1 was found to be closely linked to the skeletal muscle glycogen phosphorylase (PYGM) locus at 11q13.[39]

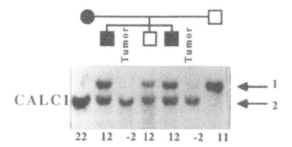

Figure 5. Loss of the wild-type allele in MEN1 associated tumors. The pedigree shows an MEN1 family with three affected members (black symbols). The autoradiogram below each family member shows the alleles for the Taq1 RFLP at the CALCA locus on chromosome 11. The genotypes with the alleles (1 and 2) for each family member are given below the autoradiogram. The affected mother is homozygous (2,2) for this marker and the father is homozygous for the alternative allele (1,1). Hence, all three children are heterozygous (1,2). Tumor tissue (insulinoma) from both affected sons (case A and B) shows loss of the paternal 1-allele.

Several subsequent studies have reported LOH for chromosome 11 markers in both sporadic and MEN1-associated parathyroid and pancreatic tumors.[70–74] While LOH at 11q13 is seen in the majority of parathyroid tumors from MEN1 patients, it is only found in one-third of the sporadic cases.[71,72,74] Several explanations for this difference are possible. It might reflect a difference in pathogenesis, or that only a subgroup of sporadic parathyroid tumors are caused by genetic alterations at the MEN1 locus. Cloning of the MEN1 gene will hopefully permit identification of minor genetic alterations (e.g., point mutations or small deletions) of the gene. Methods with higher resolution than that obtained by LOH will then be required.

Although allele losses in MEN1-associated pancreatic tumors were the key to the localization of the disease locus, similar studies on medullary thyroid carcinomas (MTC) and pheochromocytomas did not provide the information that made it possible to localize the multiple endocrine neoplasia type 2A (MEN2A) locus. Epidemiological comparisons of MTC and pheochromocytomas in MEN2A patients, and in patients with sporadic disease, suggested the necessity of two mutational events for oncogenesis.[75] After plotting the age of onset as a function of the fraction of cases not yet diagnosed, it was proposed that the MEN2A gene is a recessive tumor suppressor gene, similar to what was initially hypothesized by Knudson for the *rb* gene.[75,13]

Given the assumption that LOH involving a specific chromosome would indicate the localization of the disease locus, pheochromocytomas and MTCs were screened with highly polymorphic markers on some chromosomes.[76] In 7 of the 14 tumors, reduced signal intensity for a chromosome 1p marker was observed, while for the other markers the constitutional genotype was always retained in the tumors.

Analysis of the parental origin of the lost allele in two families showed that it was derived from the affected parent in one case.[76] However, according to the two-mutation model one would have expected that the tumor lacked the allele derived from the unaffected parent (illustrated for MEN1 in Figure 5). Therefore, this indicates that LOH for chromosome 1 does not reflect unmasking of a recessive mutation at the MEN2A locus.

When the putative MEN2A gene was localized to the centromeric part of chromosome 10,[77,78] this chromosome was tested for the occurrence of LOH. Surprisingly, analysis of several MTCs and pheochromocytomas could only detect LOH for chromosome 10 markers in a few cases.[79,80]

Before the identification of the gene responsible for MEN2A, the most likely explanation to this was that it could be difficult to delete the centromeric region of chromosome 10 without losing the whole chromosome. If two copies of another gene on chromosome 10 were essential for survival of these cells, the loss of one MEN2A gene by non-disjunction or a large deletion would be rare events. Therefore, the second event for homozygous inactivation of the MEN2A gene might instead have involved small deletions or point mutations, which were not detectable by LOH. Furthermore, the germline chromosomal abnormality could alone be responsible for the polyclonal thyroid and adrenal hyperplasias. The clonal development of MTCs and pheochromocytomas[81] from some of these cells might then involve mutations at loci other than on chromosome 10. This would still be in agreement with the epidemiological studies which had suggested the involvement of two mutational events.[75]

Since the establishment of germline mutations in the *ret* gene being responsible for MEN2A,[82,83] it is supposed that this is an activating mutation causing a product with oncogenic features. This of course explains the lack of LOH in the region of interest, but does not explain the two mutational events proposed. It therefore seems logical to expect other events, such as the inactivation (or activation) of another gene situated at a different chromosomal region causing the MEN2A phenotype.[84]

VIII. LOH IN TUMOR PROGRESSION

Genetic alterations in tumors are usually divided into the following three types:[85]

1. primary abnormalities—essential in establishing the neoplasm;
2. secondary abnormalities—important in clonal evolution and progression of the tumor; and
3. noise—random events due to genomic instability.

To differentiate between these categories it is important to study a large number of tumors, as well as tumors in different malignancy stages. For the same reason a representative part of the genome must be analyzed.

A. The Colorectal Tumor Example

The best studied example of multistep carcinogenesis is the colorectal tumor. Two features of colorectal cancer have greatly aided the recent progress in understanding its genetics. First, the majority of colorectal cancers arise from premalignant adenomatous polyps allowing the analysis of somatic genetic changes during tumor progression.[86] Second, there are several well-defined inherited syndromes that predispose to colorectal cancer in an autosomal dominant manner.[87] One of the most widely recognized predispositions imposes a striking phenotype of multiple, adenomatous colon polyps (familial adenomatous polyposis coli). This condition may also occur in combination with benign extraintestinal growths including multiple osteomas, epidermoid cysts, desmoid tumors, thyroid tumors, and retinal hypertrophy: Gardner's syndrome.[88,89]

Following the demonstration of a constitutional interstitial deletion of 5q in a mentally retarded patient with adenomatous polyposis coli and a desmoid tumor,[90] the APC locus was assigned to 5q22 by linkage analysis.[22,23] LOH was shown in about 20% of colorectal carcinomas, leading to a hypothesized tumor suppressor gene inactivation model for the disease.[91]

Vogelstein et al. performed a general search for LOH in colorectal tumors.[92] At least one polymorphic DNA marker on every non-acrocentric chromosome arm was analyzed in a large panel of tumors (i.e., allelotyping). In this study, the most frequent LOH was surprisingly not on chromosome 5q. Thus, LOH on its own would have been unsatisfactory to localize the *apc* gene. Actually, allele losses in colorectal tumors were most often seen on chromosomes 17 and 18.

By determining the minimal region of overlapping deletions, the *p53* and the *dcc* genes were shown to be the targets for the 17p and 18q deletions, respectively.[30,31] These alterations are regarded as secondary events, demonstrating that tumor suppressor genes may also be involved in tumor progression.

Based on the frequencies by which the most common mutational events were detected in the different malignancy grades, a model for progression of colorectal tumors was proposed.[93] The specificity of the mutational events was suggested to be more significant than the order (Figure 6).

B. The Glioma Example

LOH studies are often based on previous cytogenetic analyses. For example, identification of genetic alterations in gliomas started with karyotyping of the highly malignant forms. The studies revealed the most common changes to be numerical deviations, particularly gains of chromosome 7 and loss of chromosome 10. Other frequent abnormalities were deletions and translocations of 9p and 19q, and the presence of double minute chromosomes.[91] Molecular genetic studies demonstrated LOH on 9p and 10, and amplification of the epidermal growth factor receptor (EGFR) gene on 7, confirming the cytogenetic results.[95,96] However,

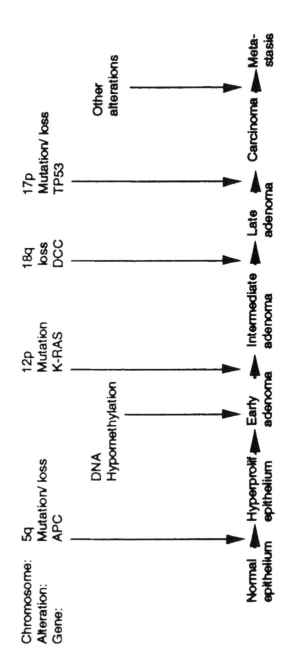

Figure 6. The Fearon and Vogelstein model for colorectal tumorigenesis.

294

nonrandom LOH was additionally found on 13 and 17.[95] For 17, the pattern of LOH was interpreted as mainly due to mitotic recombination, resulting in retention of heterozygosity at some loci, and at other loss of one allele and duplication of the remaining.[97] This type of alteration is cytogenetically not detectable.

These findings have led to the proposal of a progression sequence similar to that of the colorectal carcinoma. LOH on 17p is regarded as an early event, since it is equally frequent in all malignancy grades. Nullizygosity at the interferon gene cluster on chromosome 9p is, on the other hand, only found in the highly malignant glioblastomas, and is therefore suggested to be a late event. Somewhere between these genetic events, chromosome 10 deletions, heterozygous interferon gene deletions, *p53* mutations, and EGFR gene amplification occur.[95,96,98]

IX. DELETION MAPPING

When LOH is detected to a significant extent, an important step has been taken in mapping a putative tumor suppressor gene involved in the genetics of a certain tumor type. To further map the region of interest, the consistent step to follow is identification of the minimal region of overlapping deletions, thereby restricting the gene containing area. This is usually referred to as deletion mapping. The more polymorphic DNA markers used for this purpose, the more obvious it is that the deletion patterns are far more complex than originally suggested (see Figure 3).

For instance, a detailed deletion mapping of chromosome 10 in gliomas revealed three distinct regions with high incidence of LOH: one telomeric region on 10p, and both telomeric and centromeric locations on 10q.[99,100] However, the studies did not reveal whether translocations, multiple rearrangements on one, or different deletions on both chromosomes 10 occurred.

By analysis of the parental genotypes, these alternatives can be distinguished— an approach that was used in familial breast carcinomas. Allele losses in breast cancer mainly involves chromosomes 16 and 17.[42–44] Deletion mapping of chromosome 17 revealed two distinct regions on 17p, and two on 17q.[101] In several cases, two or more of these regions were deleted in the same tumor, while heterozygosity was maintained for other chromosome 17 loci. Determination of the parental origin of the lost alleles showed that deletions affecting these regions were independent events. The LOH shown in Figure 7 can be explained by mitotic recombination, a translocation, multiple deletions on one chromosome 17, or single deletions on both chromosomes. After parental genotyping, only the latter possibility remained.

In MEN 1-associated tumors, LOH studies revealed three regions with allele losses: additional to the MEN 1 locus on 11q13 itself, a region telomeric on 11q, and one on 11p.[102,74] In context to these observations, this particular pattern may reflect that certain regions of chromosome 11 must be maintained and are required for cell survival, or give growth advantage. One of the retained regions actually overlapped the 11q13 amplicon, frequently detected in several other tumor types.[103,104]

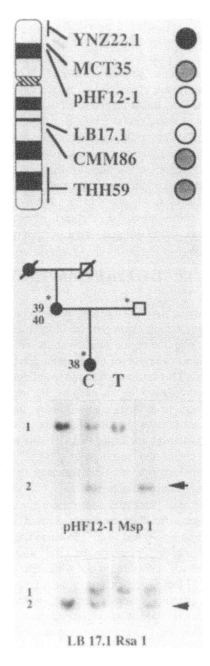

Figure 7. LOH on chromosome 17 in a case of familial breast carcinoma. RFLP analysis showed LOH (*empty circle*) at two regions flanking the centromere, and retained heterozygosity (*black circle*) at the distal part of 17p (*grey circles*: not informative). Parental genotyping showed that the deletion on 17p eliminated the paternal allele, and that the 17q deletion involved the maternal allele.

X. CONCLUSIONS: THE RESEARCHER AND THE TUMOR

To summarize, let us illustrate the advantages and disadvantages of the "loss of heterozygosity" (LOH) technique presented in this chapter by discussing a fictitious situation: You, a researcher in cancer genetics, have in your hands a collection of tumors previously investigated poorly. How do you proceed?

A common strategy does not exist. The nature of the disease, the information available, and the quality of the material will determine which methods can be applied. Considerations of this kind before starting an experiment may save both time and money.

Does the cancer exist in a heritable form (i.e., are any families available for the studies)? As an example, colon and breast carcinomas do exist in forms with Mendelian inheritance. The "sporadic" forms that appear comparatively often in the population may also comprise hereditary components: the statistical risk of an individual to acquire these cancers increases drastically with every relative affected. In fact, this may be relevant for most cancers.[87,105,106]

Linkage is sometimes hidden by heterogeneity of the disease; focusing on families with a homogenous clinical appearance may unmask this phenomenon. For example, linkage in breast cancer to chromosome 17q was established by selecting families with frequent premenopausal tumor onset.[107]

Neoplasias with strict inheritance, thus probably of monogenic origin, are thought to be rare. However, they may be more common than generally believed since they can be hidden by variations in their clinical appearance. MEN1 serves as a good example: from 130 patients in a large Tasmanian family, only 3 were diagnosed as MEN1 before kinship was recognized.[108] The skill of the investigator "asking the patients the right questions" would be crucial in such situations. A tight cooperation between the molecular researcher and the clinician is therefore highly beneficial.

Provided no additional information by related methods (see below) is available, today's method of choice for monogenic inheritable cancer is linkage analysis. On the basis of microsatellite markers, linkage is more informative and less time consuming than classical Southern analysis. In addition, linkage analysis is not restricted to genes with certain properties, while LOH is only applicable to tumor suppressor genes.

If family history reveals that your material of tumors is non-hereditary, or if the families in question are too small to provide informativity, other approaches than linkage analysis are required.

Are the samples suitable for LOH analysis? The method demands tumors of monoclonal origin, of a certain tissue homogeneity, and a minimum quality of undegraded DNA. It also requires availability of constitutional DNA from every patient. The least ambiguous results will be obtained with fresh-frozen tumors and classical RFLP or VNTR markers. Paraffin-embedded tissue DNA, irrespective of its age, is still suitable for microsatellite marker analysis. The method of preparation should be known since fixating agents cause DNA breakdown to a different extent, and thus restrict the choice of markers.[109]

Most important, again, is complete access to the clinical data. A "breast tumor" or a "thyroid tumor" is not sufficient. Age and sex of the patient, treatment, survival time, and site and stage of the tumor are essential for interpretation of the LOH results. Again, it is worth emphasizing the cooperation with clinicians and histopathologists.

Do other, faster methods give information about the locus of interest? A number of methods exist that may complement and direct LOH studies. *One good metaphase saves a hundred blots*: In spite of its limitations, classical cytogenetics gives an immediate view over gross rearrangements within the tumor genome. In addition, new techniques are presently refining cytogenetic analysis. Using fluorescently labeled markers for *in situ* hybridization (FISH), in context with confocal laser scanning microscopes, a resolution of less than 100 kilobases can be obtained.[110,111] Using chromosome specific libraries, and cosmids and yeast artificial chromosomes as markers, even minor chromosome rearrangements can be detected.[112] Single chromosomes can be amplified by degenerated oligonucleotide primer PCR.[113] Rearranged chromosomes fluorescently PCR-amplified can be hybridized against normal metaphases, thus revealing their derivation.[114]

Comparative genomic hybridization (CGH) will probably revolutionize molecular genetics, especially in cancer research. With CGH, regions of both gains and losses (deletions, duplications, amplicons) can be detected.[115,116] In short, tumor DNA and normal reference DNA, differentially labeled with fluorochromes, are hybridized simultaneously to normal metaphase chromosomes. The alterations are detected as changes in the ratio of the intensities of the fluorochromes. The method is very promising, though still in its infancy.

In spite of all newly developed techniques, LOH can not yet be replaced. The results are fairly easy to obtain and interpret, and certain genetic alterations can only be detected this way. In fact, the fascinating phenomenon of instability in the replication of dinucleotide repeats was unexpectedly discovered by LOH studies of colon carcinomas.[117,118,119]

LOH still holds its position in human cancer research.

ACKNOWLEDGMENTS

This work was supported by the Swedish Cancer Society, the Swedish Medical Research Council, the Magnus Bergwall Foundation, the Marcus Borgström Foundation, the Lars Hierta Foundation, and the Karolinska Institute.

REFERENCES

1. Mulvihill, J. J.; Miller, R. W.; Frauneni, J. F. *Genetics of Human Cancer*. Raven Press, New York, 1977.
2. McKusick, V. A. *Mendelian Inheritance in Man: Catalogs of Autosomal Dominant, Autosomal Recessive, and X-linked Phenotypes*. Johns Hopkins University Press, Baltimore, 1992.
3. Lele, K. P.; Penrose, L. S.; Stallard, H. B. Chromosome deletion in a case of retinoblastoma. Ann. Hum. Genet. **1963**, *27*, 171–174.
4. Francke, U. Retinoblastoma and chromosome 13. *Cytogenet. Cell Genet.* **1976**, *16*, 131–134.
5. Boveri, T. *Zur Frage der Entstehung Maligner Tumoren*. Verlag Gustav Fischer, Jena, 1914.
6. Caspersson, T.; Zech, L.; Johannson, C.; Modest, E. J. Identification of human chromosomes by DNA-binding fluorescing agents. *Chromosoma* **1970**, *30*, 215–227.
7. Rowley, J. D. A new consistent chromosomal abnormality in chronic myelogenous leukaemias identified by quinacrine fluorescence and Giemsa staining. *Nature* **1973**, *243*, 290–291.
8. Cairns, J. The origin of human cancers. *Nature* **1981**, *289*, 353–357.
9. Setlow, R. B. Repair deficient human disorders and cancer. *Nature* **1978**, *271*, 713–717.
10. Bishop, J. M. The molecular genetics of cancer. *Science* **1987**, *235*, 305–311.
11. Bishop, J. M. Molecular themes in oncogenesis. *Cell* **1991**, *64*, 235–248.
12. de Mars, R. 23rd Annual Symp. Fundamental Cancer Research 1969. Williams & Wilkings 1970, 105.
13. Knudson, A. G. Mutation and cancer: statistical study of retinoblastoma. *Proc. Natl. Acad. Sci. USA* **1971**, *68*, 820–823.
14. Knudson, A. G. Retinoblastoma: a prototypic hereditary neoplasm. *Semin. Oncol.* **1978**, *5*, 57–60.
15. Harris, H.; Miller, O. J.; Klein, G.; Worst, P.; Tachibana, T. Suppression of malignancy by cell fusion. *Nature* **1969**, *223*, 363–368.
16. Huang, H-J. S.; Yee, J-K.; Shew, J-Y.; et al. Suppression of the neoplastic phenotype by replacement of the *Rb* gene in human cancer cells. *Science* **1988**, *242*, 1563–1566.
17. Kinzler, K. W.; Nilbert, M. C.; Su, L-K.; et al. Identification of the FAP locus genes from chromosome 5q21. *Science* **1991**, *253*, 661–669.
18. Wallace, M. R.; Marchuk, D. A.; Andersen, L. B.; et al. Type 1 neurofibromatosis gene: identification of a large transcript disrupted in three NF1 patients. *Science* **1990**, *249*, 181–186.
19. Viskochil, D.; Buchberg, A. M.; Xu, G.; et al. Deletions and a translocation interrupt a cloned gene at the neurofibromatosis type 1 locus. *Cell* **1990**, *62*, 187–192.
20. Cawthon, R.M.; Weiss, R.; Xu, G.; et al. A major segment of the neurofibromatosis type 1 gene: cDNA sequence, genomic structure, and point mutations. *Cell* **1990**, *62*, 193–201.
21. Trofatter, J.A.M.; Rutter, J.L.; Murrel, J.R.; et al. A novel Moesin-Ezrin-Radixin-like gene is a candidate for the neurofibromatosis 2 tumor suppressor. *Cell* **1993**, *72*, 791–800.
22. Bodmer, W.F.; Bailey, C.J.; Bodmer, J.; et al. Localization of the gene for familial adenomatous polyposis on chromosome 5. *Nature* **1987**, *328*, 614–616.
23. Leppert, M.; Dobbs, M.; O'Connell, P.; et al. The gene for familial polyposis coli maps to the long arm of chromosome 5. *Science* **1987**, *238*, 1411–1413.
24. Barker, D.; Wright, E.; Nguyen, K.; et al. Gene for von Recklinghausen neurofibromatosis is in the pericentromeric region of chromosome 17. *Science* **1987**, *236*, 1100–1102.
25. Seizinger, B.R.; Rouleau, G.A.; Ozelius, L.J.; et al. Genetic linkage of von Recklinghausens neurofibromatosis to the nerve growth factor receptor gene. *Cell* **1987**, *49*, 589–594.
26. Rouleau, G.A.; Wertelecki, W.; Haines, J.L.; et al. Genetic linkage of bilateral acoustic neurofibromatosis to a DNA marker on chromosome 22. *Nature* **1987**, *329*, 5419–5423.
27. Rose, E.A.; Glaser, T.; Jones, C.; et al. Complete physical map of the WAGR region of 11p13 localizes a candidate Wilms' tumor gene. *Cell* **1990**, *60*, 495–508.
28. Call, K.M.; Glaser, T.; Ito, C.Y.; et al. Isolation and characterization of a zinc finger polypeptide gene at the human chromosome 11 Wilms' tumor locus. *Cell* **1990**, *60*, 509–520.

29. Riccardi, V. M.; Sujansky, E.; Smith, A. C.; Francke, U. Chromosomal imbalance in the aniridia-Wilms' tumor association: 11p interstitial deletion. *Pediatrics* **1978**, *61*, 604–610.
30. Fearon, E. R.; Cho, K. R.; Nigro, J. M.; et al. Identification of a chromosome 18q gene that is altered in colorectal cancers. *Science* **1990**, *247*, 49–56.
31. Baker, S. J.; Fearon, E. R.; Nigro, J. M.; et al. Chromosome 17 deletions and p53 mutations in colorectal carcinomas. *Science* **1989**, *244*, 217–221.
32. Linzer, D. I. H.; Levine, A. J. Characterization of a 54K dalton cellular SV40 tumor antigen present in SV40-transformed cells and uninfected embryonal carcinoma cells. *Cell* **1979**, *17*, 43–52.
33. Lane, D. P.; Benchimol, S. p53: oncogene or anti-oncogene. *Genes Dev.* **1990**, *4*, 1–18.
34. Vogelstein, B.; Kinzler, K. W. p53 function and dysfunction. *Cell* **1992**, *70*, 523–526.
35. Lane, D.P. p53, guardian of the genome. *Nature* **1992**, *358*, 18–19.
36. Lane, D.P. A death in the life of p53. *Nature* **1993**, *362*, 786–787.
37. Malkin, D.; Li, F.P.; Strong, L.C.; et al. Germline p53 mutations in a familial syndrome of breast cancer. *Science* **1990**, *250*, 1233–1238.
38. Srivastava, S.; Zou, Z.; Pirollo, K.; Blattner, W.; Chang, E.H. Germline transmission of a mutated *p53* gene in a cancer-prone family with Li-Fraumeni syndrome. *Nature* **1990**, *2*, 132–134.
39. Larsson, C.; Skogseid, B.; Öberg, K.; Nakamura, Y.; Nordenskjöld, M. Multiple endocrine neoplasia type 1 gene maps to chromosome 11 and is lost in insulinoma. *Nature* **1988**, *332*, 85–87.
40. Seizinger, B. R.; Martuza, R. L.; Gusella, J. F. Loss of genes on chromosome 22 in tumorigenesis of human acoustic neuroma. *Nature* **1986**, *322*, 664–667.
41. Seizinger, B. R.; De La Monte, S.; Atkins, L.; Gusella, J. F.; Martuza, R. L. Molecular genetic approach to human meningiomas: loss of genes on chromosome 22. *Proc. Natl. Acad. Sci. USA* **1987**, *84*, 5419–5423.
42. Larsson, C.; Byström, C.; Skoog, L.; Rotstein, S.; Nordenskjöld, M. Genomic alterations in human breast carcinomas. Genes Chrom. Cancer **1990**, *2*, 191–197.
43. Sato, T.; Tanigami, A.; Yamakawa, K.; et al. Cumulative allele losses promote tumor progression in primary breast cancer. *Cancer Res.* **1990**, *50*, 7184–7189.
44. Devilee, P.; van Vliet, M.; van Sloun, P.; et al. Allelotype of human breast carcinoma: a second major site for loss of heterozygosity is on chromosome 6q. *Oncogene* **1991**, *6*, 1705–1711.
45. Cavenee, W.K.; Dryja, T.P.; Phillips, R.A.; et al. Expression of recessive alleles by chromosomal mechanisms in retinoblastoma. *Nature* **1983**, *305*, 779–784.
46. Seizinger, B. R.; Klinger, H. P.; Junien, C.; et al. Report of the committee on chromosome and gene loss in human neoplasia. *Cytogenet. Cell Genet.* **1991**, *58*, 1080–1096.
47. Lasko, D.; Cavenee, W.; Nordenskjöld, M. Loss of constitutional heterozygosity in human cancer. *Ann. Rev. Genet.* **1991**, *25*, 281–314.
48. Kan, Y. W.; Dozy, A. M. Polymorphism of the DNA sequence adjacent to human β-globin structural gene: relationship to sickle cell mutation. *Proc. Natl. Acad. Sci. USA* **1978**, *75*, 5631–5635.
49. Jeffreys, A. J. DNA sequence variations in the Gγ, +Aγδ-, and β-globin genes of man. *Cell* **1980**, *18*, 1–10.
50. Jeffreys, A.; Wilson, V.; Thein, S. Hypervariable "minisatellite" regions in human DNA. *Nature* **1985**, *314*, 67–73.
51. Nakamura, Y.; Leppert, M.; O'Connell, P.; et al. Variable number of tandem repeat (VNTR) markers for human gene mapping. *Science* **1987**, *235*, 1616–1622.
52. Weber, J. L.; May, P. E. A bundant class of human DNA polymorphisms which can be typed using the polymerase chain reaction. *Am. J. Hum. Genet.* **1989**, *44*, 388–396.
53. Weber, J. L. Informativeness of human (dC-dA)n × (dG-dT)n polymorphisms. *Genomics* **1990**, *7*, 524–530.
54. Southern, E. Detection of specific sequences among DNA fragments separated by gel electrophoresis. *J. Mol. Biol.* **1975**, *98*, 503–517.

55. Williamson, R.; Bowcock, A.; Kidd, K.; et al. Report of the DNA committee and catalogues of cloned and mapped genes, markers formatted for PCR and DNA polymorphisms. *Cytogenet. Cell Genet.* **1991**, *58*, 1190–1832.

56. Slamon, D.J.; Clark, G.M. Amplification of c-erbB-2 and aggressive human breast tumors? *Science* **1987**, *240*, 1796–1798.

57. Weissenbach, J.; Gyapay, G.; Dib, C.; et al. A second-generation linkage map of the human genome. *Nature* **1992**, *359*, 794–801.

58. Balaban, G.; Gilbert, F.; Nichols, W.; Meadows, A.T.; Shield, J. Abnormalities of chromosome 13 in retinoblastoma from individuals with normal constitutional karyotypes. *Cancer Genet. Cytogenet.* **1982**, *6*, 213–221.

59. Strong, L. C.; Riccardi, V. M.; Ferrell, R. E.; Sparkes, R. S. Familial retinoblastoma and chromosome 13 deletion transmitted via an insertional translocation. *Science* **1981**, *213*, 1501–1503.

60. Sparkes, R. S.; Murphree, L. A.; Lingua, R. W.; et al. Gene for hereditary retinoblastoma assigned to human chromosome 13 by linkage to esterase D. *Science* **1983**, *219*, 971–973.

61. Benedict, W. F.; Murphree, A. L.; Banerjee, A.; Spina, C. A.; Sparkes, M. C.; Sparkes, R. S. Patient with chromosome 13 deletion: evidence that the retinoblastoma gene is a recessive cancer gene. *Science* **1983**, *219*, 973–975.

62. Dryja, T. P.; Cavenee, W. K.; White, R.; et al. Homozygosity of chromosome 13 in retinoblastoma. *N. Engl. J. Med.* **1984**, *310*, 550–553.

63. Cavenee, W. K.; Hansen, M. F.; Nordenskjöld, M.; et al. Genetic origin of mutations predisposing to retinoblastoma. *Science* **1985**, *228*, 501–503.

64. Lalande, M.; Dryja, T. P.; Schreck, R. R.; Shipley, J.; Flint, A.; Latt, S. A. Isolation of human chromosome 13-specific DNA sequences cloned from flow sorted chromosomes and potentially linked to the retinoblastoma locus. *Cancer Genet. Cytogenet.* **1984**, *13*, 283–295.

65. Friend, S. H.; Bernards, R.; Rogelj, S.; et al. A human DNA segment with properties of the gene that predisposes to retinoblastoma and osteosarcoma. *Nature* **1986**, *323*, 643–646.

66. Lee, W-H.; Bookstein, R.; Hong, F.; Young, L-J.; Shew, J-Y.; Lee, E. Y-H. P. Human retinoblastoma susceptibility gene: cloning, identification, and sequence. *Science* **1987**, *235*, 1394–1399.

67. Fung, Y-K. T.; Murphree, L. A.; T'Ang, T.; Qian, J.; Hinrichs, S. H.; Benedict, W. F. Structural evidence for the authenticity of the human retinoblastoma gene. *Science* **1987**, *236*, 1657–1661.

68. Dryja, T. P.; Rapaport, J. M.; Joyce, J. M.; Petersen, R. A. Molecular detection of deletions involving band q14 of chromosome 13 in retinoblastomas. *Proc. Natl. Acad. Sci. USA* **1986**, *83*, 7391–7394.

69. Larsson, C.; Nordenskjöld, M. Multiple endocrine neoplasia. *Cancer Surv.* **1990**, *9*, 703–723.

70. Thakker, R. V.; Bouloux, P.; Wooding, C.; et al. Association of parathyroid tumors in multiple endocrine neoplasia type 1 with loss of alleles on chromosome 11. *N. Engl. J. Med.* **1989**, *321*, 218–224.

71. Friedman, E.; Sakaguchi, K.; Bale, A. E.; et al. Clonality of parathyroid tumors in familial multiple endocrine neoplasia type 1. *N. Engl. J. Med.* **1989**, *321*, 213–218.

72. Byström, C.; Larsson, C.; Blomberg, C.; et al. Localization of the MEN1 gene to a small region within chromosome band 11q13 by deletion mapping in tumors. *Proc. Natl. Acad. Sci. USA* **1990**, *87*, 1968–1972.

73. Sawicki, M. P.; Wan, Y-J. Y.; Johnson, C. L.; et al. Loss of heterozygosity on chromosome 11 in sporadic gastrinomas. *Hum. Genet.* **1992**, *89*, 445–449.

74. Friedman, E.; De Marco, L.; Gejman, P.; et al. Allelic loss from chromosome 11 in parathyroid tumors. *Cancer Res.* **1992**, *52*, 6804–6809.

75. Jackson, C. E.; Block, M. A.; Greenawald, K. A.; Tashijan, A. H., Jr. The two-mutational-event theory in medullary thyroid carcinoma. *Am. J. Hum. Genet.* **1979**, *31*, 704–710.

76. Mathew, C. G. P.; Smith, B. A.; Thorpe, K.; et al. Deletion of genes on chromosome 1 in endocrine neoplasia. *Nature* **1987**, *328*, 524–526.

77. Mathew, C. G. P.; Chin, K. S.; Easton, D. F. et al. A linked genetic marker for multiple endocrine neoplasia type 2A on chromosome 10. *Nature* **1989**, *328*, 527–528.

78. Simpson, N. E.; Kidd, K. K.; Goodfellow, P. J.; et al. Assignment of multiple endocrine neoplasia type 2A to chromosome 10 by linkage. *Nature* **1987**, *328*, 528–530.

79. Landsvater, R. M.; Mathew, C. G. P.; Smith, B. A.; et al. Development of multiple endocrine neoplasia type 2A does not involve substantial deletions of chromosome 10. *Genomics* **1989**, *4*, 246–250.

80. Nelkin, B. D.; Nakamura, Y.; White, R. W.; et al. Low incidence of loss of chromosome 10 in sporadic and hereditary human medullary thyroid carcinoma. *Cancer Res.* **1989**, *49*, 4114–4119.

81. Baylin, S. B.; Gann, D. S.; Hsu, S. H. Clonal origin of inherited medullary thyroid carcinoma and pheochromocytoma. *Science* **1976**, *193*, 321–323.

82. Mulligan, L. M.; Kwok, J. B. J.; Healey, C. S.; et al. Germline mutations of the RET protoonco-gene in multiple endocrine neoplasia type 2A. *Nature* **1993**, *363*, 458–460.

83. Donis-Keller, H.; Dou, S.; Chi, D.; et al. Mutations in the RET proto-oncogene are associated with MEN2A and FMTC. *Hum. Mol. Genet.* **1993**, *2*, 851–856.

84. Mulligan, L. M.; Gardner, E.; Smith, B. A.; Mathew, C. G. P.; Ponder, B. A. J. Genetic events in tumour initiation and progression in multiple endocrine neoplasia type 2. *Genes Chrom. Cancer* **1993**, *6*, 166–177.

85. Heim, S.; Mitelman, F. *Cancer Cytogenetics.* Alan R. Liss, New York, 1987.

86. Tierney, R. P.; Ballantyne, G. H.; Modlin, I. M. The adenoma to carcinoma sequence. *Surg. Gyn. Obstr.* **1990**, *171*, 81–94.

87. Bishop, D. T.; Thomas, H. J. W. The genetics of colorectal cancer. *Cancer Surv.* **1990**, *9*, 585–604.

88. Gardner, E. J. A genetic and clinical study of intestinal polyposis, a predisposing factor for carcinoma of the colon and rectum. *Am. J. Hum. Genet.* **1951**, *3*, 167–176.

89. Camiel, M. C.; Mule, J. E.; Alexander, L. L.; Benninghoff, D. L. Association of thyroid carcinoma with Gardner's syndrome in siblings. *N. Engl. J. Med.* **1968**, *278*, 1056–1058.

90. Herrera, L.; Kakati, S.; Gibas, L.; Pietrzak, E.; Sandberg, A. Gardner syndrome in a man with an interstitial deletion of 5q. *Am. J. Med. Genet.* **1986**, *25*, 473–476.

91. Solomon, E.; Voss, R.; Hall, V.; et al. Chromosome 5 allele loss in human colorectal carcinomas. *Nature* **1987**, *328*, 616–619.

92. Vogelstein, B.; Fearon, E. R.; Kern, S. E.; et al. Allelotype of colorectal carcinoma. *Science* **1989**, *244*, 207–210.

93. Fearon, E. R.; Vogelstein, B. A genetic model for colorectal tumorigenesis. *Cell* **1990**, *61*, 759–767.

94. Bigner, S. H.; Mark, J.; Burger, P. C.; et al. Specific chromosomal abnormalities in malignant human gliomas. *Cancer Res.* **1988**, *48*, 405–411.

95. James, C. D.; Carlbom, E.; Dumanski, J. P.; et al. Clonal genomic alterations in glioma malignancy stages. *Cancer Res.* **1988**, *48*, 5546–5551.

96. James, C. D.; He, H.; Carlbom, E.; Nordenskjöld, M.; Cavenee, W. K.; Collins, V. P. Chromosome 9 deletion mapping reveals Interferon α and Interferon β-1 gene deletions in human glial tumors. *Cancer Res.* **1991**, *51*, 1684–1688.

97. James, C. D.; Carlbom, E.; Nordenskjöld, M.; Collins, V. P.; Cavenee, V. K. Mitotic recombination of chromosome 17 in astrocytomas. *Proc. Natl. Acad. Sci. USA* **1989**, *86*, 2858–2862.

98. Sidransky, D.; Mikkelsen, T.; Schwechheimer, K.; Rosenblum, M. L.; Cavenee, W.; Vogelstein, B. Clonal expansion of p53 mutant cells is associated with brain tumor progression. *Nature* **1992**, *355*, 846–847.

99. Rasheed, A. B. K.; Fuller, G. N.; Friedman, A. H.; Bigner, D. D.; Bigner, S. H. Loss of heterozygosity for 10q loci in human gliomas. *Genes Chrom. Cancer* **1992**, *5*, 75–82.

100. Karlbom, A. E.; James, C. D.; Boëthius, J.; et al. Loss of heterozygosity in malignant gliomas involves at least three distinct regions on chromosome 10. *Hum. Genet.* **1993**, *92*, 169–174.

101. Lindblom, A.; Skoog, L.; Ikdahl Andersen, T.; Rotstein, S.; Nordenskjöld, M.; Larsson, C. Four separate regions on chromosome 17 show loss of heterozygosity in familial breast carcinomas. *Hum. Genet.* **1993**, *91*, 6–12.
102. Larsson, C.; Weber, G.; Janson, M. Sublocalization of the multiple endocrine neoplasia type 1 gene. *Henry Ford Hosp. Med. J.* **1992**, *40*, 159–161.
103. Lammie, G. A.; Peters, G. Chromosome 11q13 abnormalities in human cancer. *Cancer Cells* **1991**, *3*, 413–420.
104. Gaudray, P.; Szepetowski, P.; Escot, C.; Birnbaum, D.; Theillet, C. DNA amplification at 11q13 in human cancer: from complexity to perplexity. *Mutat. Res.* **1992**, *276*, 317–328.
105. Lynch, H. T.; Watson, P.; Lynch, J. F. Epidemiology and risk factors. *Clin. Obst. Gyn.* **1989**, *32*, 750–760.
106. Houlston, R. S.; McCarter, E.; Parbhoo, S.; Scurr, J. H.; Slack, J. Family history and risk of breast cancer. *J. Med. Genet.* **1992**, *29*, 154–157.
107. Hall, J. M.; Lee, M. K.; Newman, B.; et al. Linkage of early-onset familial breast cancer to chromosome 17q21. *Science* **1990**, *250*, 1684–1689.
108. Shepherd, J. J. The natural history of multiple endocrine neoplasia type 1—highly uncommon or highly unrecognized? *Arch. Surg.* **1991**, *126*, 935–952.
109. Greer, C. E.; Peterson, S. L.; Kiviat, N. B.; Manos, M. M. PCR amplification from paraffin embedded tissues. *Amer. J. Clin. Path.* **1991**, *95*, 117–124.
110. Pinkel, D.; Straume, T.; Gray, J. W. Cytogenetic analysis using quantitative, high sensitivity fluorescence hybridization. *Proc. Natl. Acad. Sci. USA* **1986**, *83*, 2934–2938.
111. Wiegant, J.; Kalle, W.; Mullenders, L.; et al. High-resolution *in situ* hybridization using DNA halo preparations. *Hum. Mol. Genet.* **1992**, *1*, 587–591.
112. Cremer, T.; Lichter, P.; Berden, J.; et al. Detection of chromosome aberrations in metaphase and interphase tumor cells by *in situ* hybridization using chromosome-specific library probes. *Hum. Genet.* **1988**, *80*, 235–246.
113. Telenius, H.; Pelmear, A. H.; Tunnacliffe, A.; et al. Cytogenetic analysis by chromosome painting using DOP-PCR amplified flow-sorted chromosomes. *Genes Chrom. Cancer.* **1992**, *4*, 257–263.
114. Blennow, E.; Telenius, H.; Larsson, C.; et al. Complete characterization of a large marker chromosome by reverse and forward chromosome painting. *Hum. Genet.* **1992**, *90*, 371–374.
115. Kallioniemi, A.; Kallioniemi, O-P.; Sudar, D.; et al. Comparative genomic hybridization for molecular cytogenetic analysis of solid tumors. *Science* **1992**, *258*, 818–821.
116. du Manoir, S.; Speicher, M. R.; Joos, S.; et al. Detection of complete and partial chromosome gains and losses by comparative genomic in situ hybridization. *Hum. Genet.* **1993**, *90*, 590–610.
117. Thibodeau, S. N.; Bren, G.; Schaid, D. Microsatellite instability in cancer of the proximal colon. *Science* **1993**, *260*, 816–819.
118. Aaltonen, L. A.; Peltomäki, P.; Leach, F. S.; et al. Clues to the pathogenesis of familial colorectal cancer. *Science* **1993**, *260*, 812–816.
119. Lindblom, A.; Tannergård, P.; Werelius, B.; Nordenskjöld, M. Genetic mapping of a second locus predisposing to hereditary non-polyposis colon cancer. *Nature Genetics* **1993**, *5*, 279–282.

THE ROLE OF THE *BCR/ABL* ONCOGENE IN HUMAN LEUKEMIA

Peter A. Benn

I. INTRODUCTION

Cancer is the consequence of an accumulation of multiple critical genetic alterations that confer a growth advantage. The complexity of the process is apparent from both the diversity of genes that are involved in the process (oncogenes and tumor

Advances in Genome Biology
Volume 3B, pages 305–335.
Copyright © 1995 by JAI Press Inc.
All rights of reproduction in any form reserved.
ISBN: 1-55938-835-8

Table 1. Common Abbreviations*

Abbreviation	Alternative Forms	Brief Definition
ABL	c-*abl*, c-*ABL*	Human *ABL* gene
ABL	abl protein, P145, P145^{ABL}	Human ABL protein
c-*abl*		Normal *abl* gene (other than human or viral)
BCR	bcr, bcr gene, *phl*, *PHL*	Human *BCR* gene
BCR	bcr protein, P160 and P180 or P190, P160^{BCR} and P180^{BCR} or P190^{BCR}	Human BCR protein
M-bcr	bcr, *bcr*, bcr-1	Major breakpoint cluster region, 5.8 kb including exons 12–15 of *BCR*
m-bcr	m-bcr-1, bcr-2	Minor breakpoint cluster region, 3′ end of intron 1 of *BCR*
K-28		m-RNA with a junction of *BCR* exon 14 to *ABL* exon 2
L-6		m-RNA with a junction of BCR exon 13 to *ABL* exon 2
Ph	Ph^1, Ph′, 22q-	The Philadelphia chromosome

Note: *The abbreviations used in this review for the *BCR*, *ABL*, and fusion *BCR/ABL* genes. Some of the most common alternative forms used in other publications are also listed.

suppressor genes with diverse functions), the mechanisms of genetic alteration (amplification, overexpression, deregulation, deletion, point mutation, and translocation), and from the subtleties of the alterations arising within individual genes that can elicit a proliferative advantage. In this review, one example of an oncogene activation by translocation is considered in detail. The translocation fuses two human genes, *BCR* and *ABL,* and this juxtaposition appears to be a critical change in some leukemias. The gene fusion was the first recognized example of a specific genetic change in a human cancer. Following the cytogenetic identification of the translocation, the genetic rearrangement was characterized precisely by molecular techniques. This review discusses the role of the *BCR/ABL* fusion in the pathogenesis of human leukemia, emphasizing both the common features of the genetic alteration in different patients and the diversity that can exist within this single alteration in cancer.

Because abbreviations used by different investigators vary and have led to some confusion, a summary is provided in Table 1 of the terms used in this chapter together with same of the alternate forms commonly encountered in the literature.

II. THE PHILADELPHIA CHROMOSOME POSITIVE LEUKEMIAS

Chronic myeloid leukemia (CML) is a clonal myeloproliferative disease of a pluripotent stem cell.[1-3] The early stage of the disease is relatively benign ("chronic phase") characterized by an overproduction of granulocytes. Myeloid cells in the peripheral blood show all stages differentiation. Progressive splenomegaly and occasionally hepatomegaly occur as increasing myeloid cells are produced. The rate of cell division of myeloid cells is not increased but the life-span of the cells appears to be extended.[4] The median age of onset of the disease is approximately 50 years with similar numbers of male and female patients.[5] The duration of the chronic phase is highly variable with a median of approximately 3 years.[6]

In approximately 50% of patients with CML, the disease progresses from the chronic phase to an "accelerated phase." This is characterized by splenomegaly and leukocytosis and cells may become resistant to chemotherapy. Basophils and eosinophils may increase and thrombocytosis and myelofibrosis may develop.[8,9]

In most patients the disease progresses to an acute terminal phase, referred to as "blast crisis." Blast crisis may follow an accelerated phase or the patient may progress directly from chronic phase. Generally, blast crisis is characterized by greater than 30% blast cells in peripheral blood, bone marrow, or both.[8,10] Extramedullary blastic transformation may occur in which a localized tumor mass (myeloblastoma) is present at a lymph node or elsewhere.[11] The blast cells are myeloid in approximately 70% of cases and lymphoid in approximately 30% of cases.[12] Rare cases of biphenotypic or mixed lymphoblastic/myeloblastic cell types have also been observed.[13–15] Median survival after the onset of blast crisis is only approximately 3 to 6 months.[6] In many respects, blast crisis CML is similar to some acute leukemias.

Greater than 90% of cases of CML contain a Philadelphia chromosome (Ph—see Section III).[16] However, the chromosome is also found in 2–6% of childhood acute lymphoblastic leukemias (ALL) and 19–30% of adult ALL.[17–18] These leukemias are often phenotypically B or pre-B cell,[13] although T-cell acute leukemias may also have a Ph chromosome.[19] Ph-positive ALL may have a worse response and survival than Ph-negative ALL.[20] Morphologically, these leukemias are usually classified as L1 or L2 and are terminal deoxynucleotidyl transferase (TdT) positive.[21] At least some of the patients noted with Ph-positive ALL might be considered as patients with lymphoid transformation of CML with a short or unrecognized chronic phase.[22] However, based on clinical response and relapse criteria, as well as some ancillary cytomorphologic criteria, some patients with Ph-positive ALL appear to be quite distinct from blast crisis CML patients and are much more typical of other ALL patients.[23] Rare cases of AML are also Ph positive.[24] These leukemias have been typed morphologically as M1, M2, and M4.[25]

The induction of all types of leukemia appears to involve both genetic and environmental factors.[5,26] The association between radiation exposure and leukemia incidence is well established, and it is likely that other agents which cause chromosome breakage are leukemogenic. It is, therefore, perhaps not surprising that specific chromosome rearrangements are implicated in the development of leukemias.

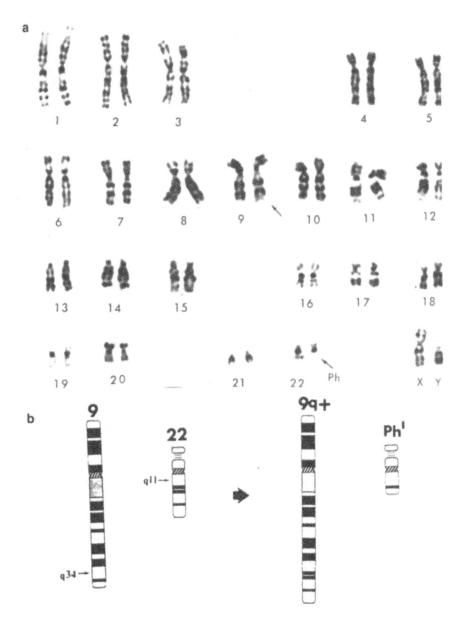

Figure 1. (a) Karyotype prepared from the bone marrow cells of a patient with CML. The Philadelphia (Ph) chromosome and derivative chromosome 9 (9q+) are identified by arrows. (b) Diagram illustrating the reciprocal translocation t(9;22)(q34;q11) that gives rise to the Ph chromosome and 9q+ chromosome.

308

III. CYTOGENETICS

In 1960, Nowell and Hungerford noted that a specific marker chromosome was present CML.[27] The marker chromosome was named the Philadelphia chromosome (Ph) after the city of its discovery. The Ph chromosome was noted to be present in approximately 90% of cases of CML.[16] Following the introduction of chromosome banding, it was observed that the Ph chromosome was usually the result of an apparently balanced reciprocal translocation between a chromosome 9 and a chromosome 22.[28] The translocation involves band q34.1 of chromosome 9 and band q11.21 on chromosome 22 (Figure 1).[29]

Approximately 4% of CML patients show "complex" exchanges involving three or more chromosomes.[30,31] In such cases, chromosome 9 and 22 are still involved and the Ph chromosome appears to be identical to that seen in simple t(9;22) translocations. In a further 3–4% of patients, chromosome 9 does not appear to be obviously involved although the Ph chromosome again appears to be identical to that seen with simple t(9;22) translocations.[32] This latter group is sometimes referred to as "simple variant" translocations.[32] "Masking" of Ph chromosome by additional chromosomal exchanges also sometimes occurs.[33] Although some cases diagnosed as Ph chromosome negative CML are in fact other hematologic disorders, it is clear that Ph-negative CML does exist.[34] In at least some of these cases, *BCR* and *ABL* are juxtaposed with the exchange occurring below the level of resolution used by light microscopy.[35-37]

Additional cytogenetic changes are frequently seen in CML, particularly during blast crisis.[16] These additional changes are nonrandom in nature (Table 2) with considerable diversity in the karyotype of blast crisis CML from patient to patient.

Table 2. Additional Chromosome Abnormalities in CML[*]

	Chronic Phase	*Blast Crisis*
+Ph	3	32
+8	3	29
i(17q)	0	13
+19	1	14
+21	3	8
–Y	3	2
Other	5	25
Any	11	67

Note: [*]Percentage of cases of CML with additional cytogenetic changes in chronic phase and blast crisis. Based on data in Ref. 197.

Clonal evolution probably also involves additional genetic changes such as p53[38,39] and *ras* gene mutation[40,41] not detectable by cytogenetic analysis.

The Ph chromosome in Ph-positive ALL is indistinguishable cytogenetically from that in CML.[17–18] Like CML, complex and variant translocations have been observed in Ph-positive ALL.[42] However, Ph-positive ALL patients may more frequently show some normal metaphases in their bone marrow cells at diagnosis.[43] Another difference between Ph-positive ALL and CML is that nonrandom secondary cytogenetic changes (extra Ph, +8, i(17q), and +19) frequently seen in blast crisis CML are often not present in ALL.

As will be discussed in detail later, a further distinction between Ph-positive CML and some Ph-positive ALL can be seen at the molecular level in terms of breakpoint locations. Before discussing *BCR/ABL* fusions, the normal structure and function of these two genes is reviewed.

IV. THE *ABL* GENE

Before considering the structure and function of the human *ABL* gene, a brief review of the transforming v-*abl* gene of Abelson murine leukemia virus (AMuLV) is necessary.[44] A number of comprehensive reviews of A-MuLV have been published in recent years[45–48] and only a brief discussion of the virus will be included here.

AMuLV is a retrovirus that is capable of rapidly inducing lymphomas in mice.[44] The virus is also capable of transforming mouse fibroblasts and lymphoid cells *in vitro*.[49–51] The transforming property of A-MuLV is associated with the v-*abl* viral gene that encodes a tyrosine kinase.[52,53] Tyrosine kinases catalyze the transfer of the terminal phosphate of adenosine triphosphate to the hydroxyl group of tyrosine. A number of other oncogenes are tyrosine protein kinases as are several growth factor receptors.[54]

A-MuLV is thought to have originally arisen as a result of a recombination event between the less oncogenic Moloney murine leukemia virus (M-MuLV) and the mouse c-*abl* gene.[44,55,56] As a result of the recombination, the v-*abl* gene contains a 5' sequence coding for a gag viral structural protein moiety and a 3' sequence coding for part of c-*abl*. The gag viral protein contains a myristylation signal that probably directs the v-abl protein to the membrane.[57] Control of A-MuLV v-*abl* expression is mediated by promoter and enhancer signals originating from the virus.[48]

A second viral form of v-*abl* is found in the feline Hardy–Zuckerman (HZ-2) virus.[58] HZ-2 was originally isolated from a fibrosarcoma in a Siamese cat. In HZ-2 the v-*abl* is the result of a recombination between the viral *gag* gene and the feline c-*abl*. Precise recombination sites differ in the two viruses with involvement of a second feline virus gene, *pol* at the 3' recombination site in HZ-2.[59]

The human gene homolog to the mouse c-*abl* gene is referred to as *ABL*. In contrast to v-abl, the normal cellular mouse c-abl and human ABL proteins have

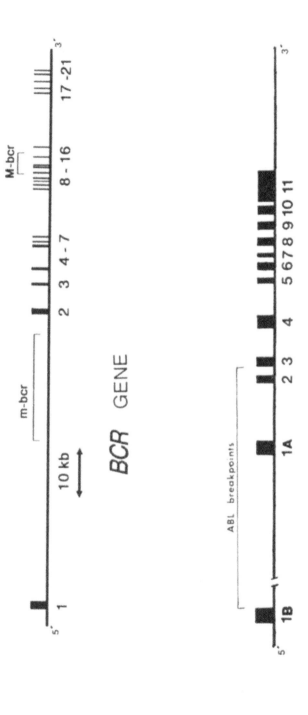

Figure 2. Diagram of the *BCR* gene (chromosome 22) and *ABL* gene (chromosome 9) showing the location of the major breakpoint cluster region (M-bcr), minor breakpoint cluster region (m-bcr), and the position of the *ABL* breakpoints.

only weak tyrosine kinase activities.[60–61] The human *ABL* gene has been extensively characterized. The *ABL* gene is at least 230 kilobases (kb) in size[62] with its 5′ and towards the centromere of chromosome 9, at band q34 (Figure 2).[63] There are 12 exons with two alternate forms of the 5′ most exon.[64] Exons 2–12 are sometimes referred to as the "common" exons since they are present in all mRNA'S transcribed from *ABL*. The alternative forms of exon 1 are 1A (located approximately 19 kb 5′ of exon 2) and 1B (located at least 200 kb 5′ of exon 1A). The substantial distance between these two alternate exons is unusual and the mechanisms for transcription over such distances are unclear.[62] The two types of mRNA's produced by alternate splicing are identifiable by their size difference (6.0 kb for exon 1A usage and 7.0 kb for exon 1B) (Table 3).[64] Based on studies on mice, alternate forms of mRNA can be found in a wide variety of tissues, but particularly in testis, thymus, and other lymphoid tissues. There is some evidence for variation in the relative amount of each type of transcript in different tissues.[65–67]

The proteins produced by human *ABL* have a molecular mass of approximately 145 kDa and are referred to as p145.[68] The tyrosine kinase function of the protein is defined by the so-called SH_1 domain (src homologous domain 1),[54] and it is this region of the protein that shows the strongest homology between species.[64] The majority of this region is essential for the transforming capability of v-abl.[69] The SH_2 domain includes amino acid sequences found in GTPase-activating proteins (GAPs) and other proteins that may be involved in signaling pathways.[48] The SH_3 domain contains sequences shared with other oncogenes, although the presence of the entire SH_3 domain does not appear to be essential for the transformation properties of the protein.[48] In fact, the SH_3 domain is a negative regulator of the kinase activity of the abl protein. Deletion of the SH_3 domain in mouse c-*abl* results in activation of the transformation ability of the protein.[70] The amino acid terminus of the protein is the end where substitutions of gag are found in v-*abl* and where BCR is juxtaposed in human leukemias (see Section VI). The carboxyl terminus does not appear to be required for kinase activity although the region does show a relatively highly conserved sequence.[48]

Table 3. Transcripts and Translation Products*

Gene	mRNA	Protein
BCR	4.5 kb, 6.0 kb	P160, P180 or P190
ABL	6.0 kb, 7.0 kb	P145
BCR/ABL (CML)	8.5 kb (K-28), 8.5 kb (L-6)	P210
BCR/ABL (ALL, AML)	8.5 kb (K-28), 8.5 kb (L-6)	P210
	7.0 kb	P190

Note: *Summary of the transcription (mRNA) and translation (protein) products for the various *BCR, ABL,* and *BCR/ABL* genes. Sizes of the mRNA and proteins are approximate and different authors cite somewhat variable numbers in their designations of these products.

Despite the large amount of information available relating to the human *ABL* gene, its normal function in cells is unknown. Among the other well-analyzed tyrosine kinases, *ABL* is most similar to the oncogene *FES*.[54] Another gene, *ARG* (*ABL*-related gene) has been described that shows substantial homology to *ABL*.[71] *ARG* is located on chromosome 1, band q24 or q25. This observation is of interest since this region has been associated with specific chromosome rearrangements in cancer.[72] This includes a specific translocation t(1;6)(q23-q25; p21-25) noted in occasional cases of myeloproliferative disorder.[73,74] While the precise details of the functions of *ARG* other putative tyrosine kinases are unclear, the insights gained from the study of *ABL* activation may be extremely helpful in unravelling parallel pathways to neoplasia.

V. THE *BCR* GENE

The *BCR* gene is located on chromosome 22 with the 5′ end of the gene closest to the centromere (see Figure 2).[75] The gene was named following the observation that the breakpoints on chromosome 22 in Ph-positive CML were clustered within a limited region, the breakpoint cluster region, or M-bcr.[76] The *BCR* gene is also frequently referred to as *phl*.

The normal *BCR* gene is approximately 130 kb in size with approximately 21 exons.[75,77,78] Like *ABL*, there is a large separation between exons 1 and 2.[79] In the case of *BCR* this distance is approximately 68 kb. The gene appears to be evolutionally conserved with substantial homology between mouse, chick, and human sequences.[80] Two RNA transcripts are made (4.5 and 6.0 kb in size)[81] with two BCR protein products of molecular mass approximately 160 and 180 kDa, or 190 kDa (see Table 3).[82,83]

BCR transcripts are detectable in human fibroblasts, B and T lymphoid, myeloid, and erythroid cell lineages.[75,77,80] The protein does not have tyrosine kinase activity (in contrast to ABL). In immunoprecipitates of BCR, a serine kinase activity has been detected[83] but this may represent a contaminant since there is no homology between BCR and any known serine kinase gene.

Sequence analysis of the *BCR* gene indicates that there is homology between BCR and GTPase-activating proteins (GAPs), and BCR appears to be a member of this interesting group of polypeptides.[84] GAPs accelerate the intrinsic rate of GTP hydrolysis of Ras-related proteins leading to down regulation of the active GTP-bound form.[85] The *ras* oncogenes regulate many processes of eukaryote cells including cell growth, cytoskeletal organization, transportation, and secretion. GAPs may act as tumor suppressor genes limiting the level of Ras oncoprotein activities.[8] It has been shown that the Ras-related protein Rac has increased GTPase-activity in the presence of BCR.[84] It is the carboxyl terminal domain of BCR that appears to have this GAP characteristic. Although the function of Rac is unknown, its homology with other Ras oncoproteins imply an important role in cellular function. The *rac* messenger RNA is present in a wide variety of cell types and there is some evidence the levels may be higher in myeloid cell lineages.[87]

Thus BCR appears to a regulatory factor for Rac, a protein that can be speculated as playing an important role in cell growth or differentiation of eukaryote cells. Another closely related GAP, *n*-chimerin exists which appears to be related to BCR.[84] However *n*-chimerin has not been implicated in fusion with ABL.

BCR shows homology with three other loci (*BCR2*, *BCR3*, and *BCR4*) which are all closely linked with *BCR* on chromosome 22.[88] It is not known what is the function of these additional related genes or even whether they are expressed.

VI. *BCR/ABL* FUSIONS IN CML

Following the localization of *ABL* chromosome 9,[89] it was established that in CML a substantial part of *ABL* is translocated to the Ph chromosome.[63,90,91] Breakage on chromosome 22 was clustered within a "breakpoint cluster region"[76] and, as a result of the genetic rearrangement, a novel mRNA species was produced,[81,92–95] with an associated fusion protein.[83,96–97] These studies unequivocally established the important mechanism of gene fusion in the pathway to neoplasia.

The breakpoint cluster region (M-bcr) of chromosome 22 was originally defined as a 5.8-kb segment that included four exons of the *BCR* gene.[76] This region contains nearly all the chromosome 22 breakpoints in patients with CML. The four exons are often referred to as M-bcr exons 1 to 4 and correspond to exons 12 to 15 of the *BCR* gene, based on the characterization of the gene by Heisterkamp et al.[79]

Within the M-bcr, the vast majority of breakpoints occur between exon 13 and exon 15.[98] Localization of breakpoints can be achieved by Southern blot analysis with combinations of multiple restriction enzyme digests of leukemic cell DNA and using multiple M-bcr probes (Figure 3).[99–100] Information on breakpoints may also be deduced by a polymerase chain reaction (PCR) technique amplifying cDNA produced from *BCR/ABL* transcripts,[101–104] although interpretation is subject to modification as a result of alternate splicing of mRNA.[105] In the PCR technique, primers to a *BCR* exon sequence and an *ABL* exon sequence are used to selectively amplify cDNA sequences where these two exons are juxtaposed (Figure 4). Both techniques indicate that breakpoints are clustered on either side of exon 14. Rare cases have been assigned breakpoints that are 5′ of exon 13,[106–108] but at least some of these may be due to technical problems associated to the assignment of breakpoints.[109] Breakpoints 3′ of exon 15 also seem to be rare.[107,110–112] Mills et al.[113] have pointed out that sequence data indicate that an out-of-frame mRNA with an early stop codon would result from a splice between *BCR* exon 15 and *ABL* exon 2. This is in contrast to the in-frame mRNAs produced as a result of splicing between *BCR* exon 13 or 14 and *ABL* exon 2. Thus, in the absence of additional mutation, the sequence and the requirement for a biologically active BCR/ABL protein product appear to define permissible breakpoints.

(a)

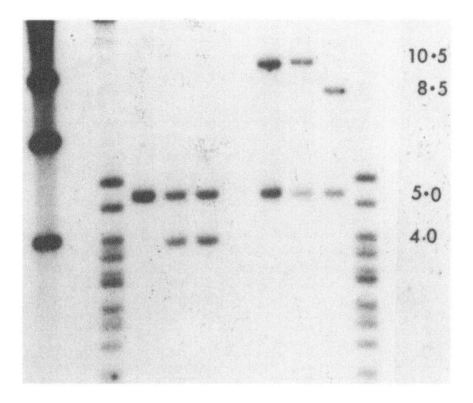

Figure 3. (a) Results of a Southern blot analysis for M-bcr rearrangement using a 1.2 kb *Hind* III/*Bgl* II probe on DNA digested with the restriction enzyme *Bgl* II. Lane 1 shows lambda DNA digested with *Hind* III. Lane 2 and 9 show molecular weight markers. Lane 3 is a control specimen; Lanes 4 and 5 are blood and bone marrow from one patient; Lanes 6 and 7 are blood and bone marrow from a second patient and Lane 8 is blood from a third patient. Each patient shows its own characteristic rearrangement autoradiograph band (in addition to the normal 5.0 kb unrearranged M-bcr band). The rearrangement band is indicative of a change in the distance between restriction enzyme sites as a result of translocation. (b) Same as (a) but with DNA digested with the restriction enzyme *Hind* III. (c) Restriction enzyme map of the M-*bcr* with the location of the restriction enzyme sites marked. From the results in (a) and (b) the location of breakpoint can be deduced. For example, patient 1 shows rearrangement with *Bgl* II digestion but not with *Hind* III. Thus, the breakpoint is likely to be between the two *Bgl* II sites shown but also 5' of the second *Hind* III site. (*continued*)

315

(b)

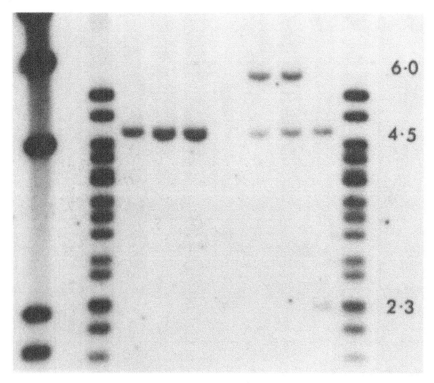

(c)

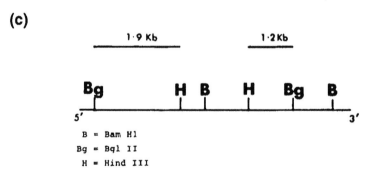

Figure 3. (continued)

In contrast to the tightly clustered range of breakpoints seen on chromosome 22 in CML, the breakpoints on chromosome 9 are to be found in a substantial region spanning 200 kb or more.[62] It was originally thought that breakpoints were always 5′ of *ABL* exon 2, the first common exon, and that the breakpoints could be 5′ of exon 1A or exon 1B. Breakpoints appear to be mostly between exon 1B and exon

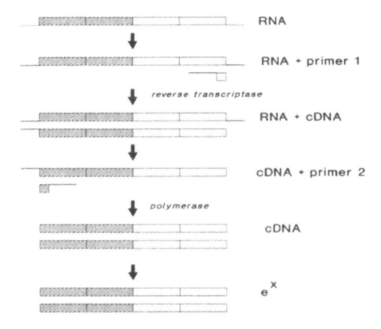

Figure 4. Diagrammatic illustration of the PCR technique for the detection of *bcr* rearrangement. In this approach RNA is extracted from cells and incubated in the presence of an oligonucleotide primer (primer 1) and the enzyme reverse transcriptase. This produces a strand of cDNA. Synthesis from this cDNA can be achieved using a second primer (primer 2) and DNA polymerase. The number of copies of the two strands of cDNA can be increased exponentially (e^x) by repeated dissociation of the double stranded DNA and repeated rounds of synthesis.

1A and may be clustered in three regions: 30 ± 5 kb, 100 ± 13 kb, and 135 ± 8 kb downstream from exon,1B.[114] Recently, a case of CML has been described with a chromosome of breakpoint between *ABL* exon 2 and *ABL* exon 3.[115] Similar observations have been made for patients with ALL.[196] It is of interest to note that these downstream breakpoints result in the removal of 17 amino acids encoded by exon 2 in the BCR/ABL fusion protein. The 17 amino acids correspond to part of the SH_3 domain of ABL, and since this domain is thought to have a negative regulatory effect on the kinase (SH_1) domain,[78] the change may well have biological significance. The frequency of these 3′ breakpoints remains to be established.

Although the translocation t(9;22)(q34;q22) appears to be balanced by cytogenetic analysis, molecular analyses indicate that there are often substantial deletions of sequences at the breakpoints.[116] Using Southern blot analyses with M-bcr probes, Popenoe et al.[116] noted that deletions can be detected in 10–20% of CML patients. Sequence analysis also indicates that deletions are present[75] and these probably occur for both chromosome 9 and chromosome 22. Additional sequences of undetermined origin may also be present at the breakpoints in the derivative chromosomes.[75] While there is a requirement for in-frame mRNA produced by the biologically significant *BCR/ABL* fusion chromosome (the Ph chromosome), no such constraint exists for the reciprocal rearrangement chromosome (the 9q+ chromosome). In fact, no transcripts are consistently detected from the *ABL/BCR* fusion on 9q+.

No substantial homology between *BCR* and *ABL* sequences have been reported despite the fact that both the carboxyl terminal of BCR and the SH3 domain of ABL have sequences associated with GAPs.[48,84] At least among newly diagnosed CML, breakpoints in the exon 13 to exon 15 segment of *BCR* appear to be random, although the number of cases studied is small.[117] An Alu repeat sequence is present between *BCR* exon 14 and exon 15,[75,118] and Alu sequences are also found within *ABL*.[119–121] However, based on the analysis of the sequences at the t(9;22) breakpoints in patients with CML, there is no evidence for sequence-specific recombination involving Alu or other[122] common repeat sequences.[75] Unlike some other tumor-associated specific chromosome rearrangements, there do not appear to be any known recombinase-specific sequences present in the vicinity of breakpoints to account for t(9;22) exchanges.

An additional consideration in evaluating the consequences of different breakpoints on chromosome 9 and 22 is the alternate splicing of mRNA transcribed from the *BCR/ABL* fusion gene.[105] Alternate splicing can sometimes occur around *BCR* exon 14 in patients with a 3′ breakpoint. That is, if a patient has a *BCR/ABL* rearrangement in which the chromosome 22 breakpoint is downstream of exon 14, the cells may contain *BCR/ABL* mRNAs with or without *BCR* exon 14 encoded sequences, or may have both species of mRNA. The two types of mRNA that are usually produced are sometimes referred to as the K28 type with *BCR* exon 14 sequence linked to *ABL* exon 2 sequence and L-6 when there is a *BCR* exon 13 to *ABL* exon 2 junction. Early studies using an RNase protection assay indicated that alternate splicing was common, based on the presence of both species of mRNA in a high proportion of patients.[123] Data from PCR amplification of cDNA (see Fig. 4) has confirmed that alternate splicing takes place although the frequency of this may be less than previously thought (see Section VIII).[101–105,124]

The extent to which other alternate splicing or incomplete splicing may occur is not well established. Romero et al.[125] studied the transcripts produced by the cell line KBM-5 (established from a CML patient with a myeloid blast crisis) and observed that two alternative BCR exon 1 sequences were sometimes utilized. The transcripts were also detected in two other blast crisis CML patient samples, but not in 12 other CML patients of various clinical stages or in six acute leukemias.

Whether or not variation in breakpoint or splicing has a significant effect on the properties of the BCR/ABL fusion protein is also unknown. The presence or absence of *BCR* exon 14 corresponds to 25 amino acids in the BCR/ABL fusion protein and the absence of *ABL* exon 2 sequence deletes 58 amino acids from

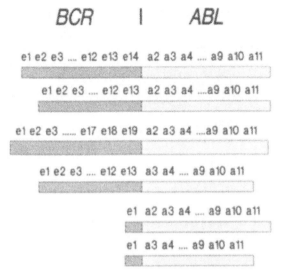

Figure 5. Some of the types of mRNA observed in the cells of patients with *BCR/ABL* gene fusions. e₁ . . . . denotes *BCR* exons and a₂ . . . . denotes *ABL* exons. The first two types of mRNA shown are the most common and are found in nearly all cases of CML and approximately 50% of Ph chromosome positive acute leukemias. The third type of mRNA is seen in rare cases of CML.[195] The fourth mRNA lacks ABL common exon 2 and has been observed in CML and ALL.[115,196] The fifth and sixth mRNAs appear to be largely confined to acute leukemias.

BCR/ABL.[75,77,78,126] The novel transcripts described by Romero et al.[125] with alternate *BCR* exon 1 sequences result in even larger changes to BCR/ABL protein structure. Two patients with CML have been described with breakage 5′ of BCR exon 19 resulting in an in-frame fusion mRNA 540 kb longer than usual and a resulting protein with an additional 180 amino acids (Figure 5).[195] Unfortunately, sensitive quantitative assays for BCR/ABL protein amount and activity are not yet available. While the BCR/ABL protein can be isolated by immunoprecipitation, protease activity or inhibition of the kinase activity have made quantitation problematical.[97,109,127,128] The proteins isolated by immunoprecipitation have a molecular mass of approximately 210,000 kDa, but minor variations are not easily detectable.[83,96,97] These proteins are referred to as P210. Van Denderen et al.[129] have described the production of a polyclonal antiserum that will specifically recognize the junction site of a BCR exon 13/ABL exon 2 fusion protein. The antiserum did not react with BCR exon 14/ABL exon 2 fusion protein. The production of antibody specific to particular types of fusions may be extremely useful in increasing an understanding of the effects of the various types of BCR/ABL proteins, as well as being potentially useful diagnostic reagents.[130]

Variation in breakpoints and transcripts does not pose a substantial problem for the routine diagnosis of CML by Southern blot analysis[100,131] or PCR.[101-104] Some myeloproliferative syndromes are clinically difficult to distinguish from CML and detection of M-bcr rearrangement can confirm a diagnosis of CML.[132] A small proportion of cases of CML show an apparently normal karyotype, yet do have M-bcr rearrangement with rearrangement involving chromosomal segments too small to be identified cytogenetically.[35,36] Other M-bcr rearrangement positive cases show complex translocations without obvious involvement of chromosome 22 or with exchanges that mask the Ph chromosome.[36] Since Ph chromosome negative CML has previously been regarded as a subgroup of patients with poorer prognosis, additional molecular characterization in this group of patients is required to reevaluate this issue.[133] Despite the reports that (atypical) M-bcr rearrangement negative CML exists,[132,134] Dr. J. Rowley has proposed that the *sine qua non* for CML is the juxtaposition of *BCR* and *ABL*.[135]

In interpreting diagnostic tests that utilize Southern blot analyses with bcr probes, rare polymorphisms need to be considered since these can lead to false positive results.[136-139] Quantitation of autoradiograph band intensities can be useful for monitoring disease[140] and residual disease following bone marrow transplantation can be detected using the PCR technique.[141] Occasional cases of CML may give a false negative result in PCR analysis due to degradation of mRNA, expression of unusual transcripts, or for other unknown reasons.[104] *In situ* hybridization techniques that take advantage of the juxtaposition of *BCR* and *ABL* may also be useful in the future for diagnosis of CML.[142]

VII. *BCR/ABL* FUSIONS IN ACUTE LEUKEMIAS

Approximately 50% of patients with Ph-positive ALL have chromosome 22 breakpoints within the M-bcr.[143,144] The remaining 50% of patients have breakpoints within the first intron of the *BCR* gene.[145,146] Surprisingly, breakage is nonrandomly distributed in this region and breakpoints are clustered to a segment referred to as m-bcr.[145,148] This is an unexpected finding since breakage anywhere throughout the intron might be expected to result in the same *BCR/ABL* mRNA. Furthermore, no specificity is apparent in the location of the breakpoints within the introns in CML (see above). Heisterkamp et al.[146] observed that six out of six breakpoints in ALL were at the 3' end of the first intron, a region consisting of about 35 kb. Denny et al.[147] noted that for eight patients studied with *BCR* intron 1 breakpoints, all were within a 20-kb 3' segment of intron 1, while Chen et al.[148] noted that six of seven patients had breakpoints within a 10.8-kb segment that they termed bcr-2. Combining these results, it would appear that there is a distinct clustering of breakpoints, although there may not be an exclusive precisely-defined region of BCR intron 1 that is involved. Chromosome 9 breakpoints appear to be similar to that seen in

CML, with most cases showing involvement of the region 5' of exon 2 and rare cases showing breakage between exon 2 and exon 3 of *ABL*.[196]

The reason for preferential location of breakpoints within BCR intron 1 has been the subject of some speculation. Denny et al.[147] have pointed out that *BCR* intron 1 contains a 1-kb deletion polymorphism with approximately one-third of chromosomes 22 lacking the 1-kb sequence that includes Alu repeats. They proposed that this sequence may in some way result in increased susceptibility toward rearrangement. Although breakage appears to be near or within Alu sequences in the few cases where breakpoints have been sequenced, the exchange does not necessarily involve recombination between homologous sequences.[119-221] Papandopoulos et al.[121] proposed that Alu or other common sequences may represent potential binding sites for proteins involved in DNA cleavage or proteins that alter chromatin structure in a way that favors cleavage. Alternatively, for reasons unknown at this time, the retention of a large *BCR* intron 1 sequence may be essential for *BCR/ABL* functional protein and that the clustering of breakpoints at the 3' end of BCR intron 1 could reflect the need to maintain specific intron sequences. In view of the widespread distribution of Alu sequences in the human genome, the location of these within the *BCR* and *ABL* genes could be coincidental.

Breakage within *BCR* intron 1 and joining to *ABL* results in a mRNA of 7.0 kb.[145,150-153] The M-bcr and m-bcr breakpoints are consistently found in Ph-positive acute leukemia but are not usually present in Ph-negative acute leukemia.[146] Ph-negative acute leukemia is not, in general, therefore attributable to a "masking" of the genetic exchange or to rearrangement of minute sized segments of chromosomes as characterizes many cases of Ph-negative CML. Thus, the acute leukemias appear to reflect a heterogenous group of disorders that do not all have a *BCR/ABL* gene fusion as a component of their etiology.

It is not known whether there is any distinct clinical differences between acute leukemias with *BCR/ABL* rearrangements involving M-bcr and those involving m-bcr. Based on a small number of patients, Secker-Walker et al.[154] could not identify any clear distinction between the two types of Ph-positive acute lymphoblastic leukemia patients. They noted that M-bcr rearrangement was sometimes confined to the lymphoblastic component of marrow or blood and suggested that heterogeneity in the translocation target cell in terms of its differentiation potential could be important in determining disease outcome. Because of the rarity of Ph-positive acute myeloid leukemia, relatively few cases have been studied.[98] Subclassification of additional acute leukemias on the basis of their *BCR* breakpoints is required and such studies could well help better delineate disease subtypes.

VIII. THE ROLE OF *BCR/ABL* IN BLAST CRISIS CML

Since each case of CML has slightly different breakpoints on chromosomes 9 and 22, the size of the genomic *BCR/ABL* junction fragments differ for each fusion (see

Fig. 3). Serial studies on patients indicate that, in general, no alteration in break-points occurs as patients progress from chronic phase through blast crisis.[107,110,111] This appears to be the case even when cytogenetic studies indicate that a second Ph chromosome has arisen during clonal evolution.[110,155] The second Ph chromo-some would therefore represent a duplication of the initial Ph chromosome rather than the product of additional gene rearrangement.

For patients sequentially studied during disease progression, a number of excep-tional or unusual situations have been described. Bartram et al.[156] reported a case of CML in which blast crisis was characterized by two Ph chromosomes and the presence of an additional M-bcr rearrangement that was associated with a novel 10.3-kb mRNA. There are at least two reports of cases of CML in which there appeared to be a loss of BCR/ABL fusion DNA.[157,158] Cytogenetic evidence for loss of the Ph chromosome has been presented but such cases seem to be rare.[159] Retention of the normal BCR allele is also the usual situation although one case has been described in which a progressive loss of germline BCR DNA was observed as the patient progressed to blast crisis.[160]

There is controversy regarding reports that some BCR/ABL fusions may be associated with shorter chronic phase durations than other fusions. A number of studies have suggested that the median chronic phase in the population of patients with a breakpoint at the 5' end of M-bcr is longer than the median chronic phase durations for patients with 3' M-bcr breakpoints.[110,111,113,117,118] Other studies have failed to confirm this trend[107,161-166] and it is clear that there are many exceptional patients who may have 5' breakpoints and short chronic-phase durations as well as patients with 3' breakpoints and unusually long chronic phases.[167,168] Heterogeneity in the populations and ascertainment biases within the patient studied clearly exist; there is a need for careful prospective studies to resolve this question.[169,170] A more rapid disease course in the population of patients with 3' M-bcr breakpoints relative to those with 5' M-bcr breakpoints could account for the relatively large number of 3' breakpoints seen in patients with blast crisis CML.[110,113,171]

It has been suggested that alternate splicing of BCR/ABL may occur more frequently during blast crisis compared to during the chronic phase.[172] Lee et al.[172] observed the simultaneous presence of the two mRNA species corresponding to the presence or absence of BCR exon 14 sequences (K-28 and L-6 junctions) in 3 of 9 blast crisis patients compared to only 2 of 21 chronic phase patients. Morgan et al.[173] observed a less striking difference in the frequency of dual expression of transcripts: 3 of 9 chronic phase patients compared to 5 of 11 blast crisis patients. Since M-bcr breakpoints were not given for the patients, it is not clear what proportion of patients had the potential for dual expression and whether the observed differences are in fact confirming that 3' breakpoints tend to be relatively more common in blast crisis CML patients. Interestingly, Lee et al.[174] have presented preliminary data that many patients with expression of an mRNA that includes BCR exon 14 (K-28 junctions) have higher platelet and white cell counts and do not respond as well to interferon therapy compared to patients with mRNA lacking BCR exon 14 (L-6 type).

In general, progression to blast crisis is not associated with a switch to production of the P190 protein that is associated with acute leukemias. Hooberman et al.[104] described a case of ALL that expressed both the mRNA associated with M-bcr rearrangement positive CML and the mRNA characteristic of M-bcr rearrangement negative, Ph-positive ALL. They speculated that the acquisition of the ALL-type of transcript might be one mechanism whereby acute lymphoid transformation can occur. In a series of 37 accelerated and blast crisis CML patients, dual expression of mRNA encoding for P210 and P190 was observed only in 3 cases.[175] In two of these three cases samples studied at the time of initial chronic phase diagnosis, only mRNA corresponding to P210 was expressed. Expression of the mRNA corresponding to P190 in blast crisis CML does not appear to necessarily correspond to the acquisition of a second Ph chromosome.[175]

The level of expression of the various fusion genes may also change during disease progression. One of the most common additional cytogenetic abnormalities that characterizes blast crisis is the acquisition of a second Ph chromosome[16] and resulting increased levels of BCR/ABL fusion protein could possibly contribute to the more aggressive leukemic cell phenotype. In the cell line K562, originally derived from a patient with blast crisis CML,[176] the BCR/ABL rearrangement is amplified four- to eightfold with corresponding enhanced expression of mRNA.[177,178] Furthermore, the level of expression of the BCR/ABL mRNA appears to be higher for some other cell lines established from acute phase CML patients.[81,92] Collins and Groudine[155] have described a patient who presented in lymphoid blast crisis with amplification of the rearranged *BCR/ABL* gene in both blast cells and granulocytes and with overexpression of *BCR/ABL* transcripts in blast cells relative to granulocytes. Weinstein et al.[36] described a patient with Ph-negative CML who had a BCR/ABL rearrangement that appeared to be duplicated (relative to the copy number for the unrearranged BCR gene) and who rapidly progressed to blast crisis. In a series of 10 patients with blast crisis CML, Andrews and Collins[177] observed significantly elevated levels of transcription of *BCR/ABL* in 4 patients. No elevation in the transcription of *BCR/ABL* was observed in a further 7 chronic phase patients. These observations suggest, but do not prove, that the levels of expression of the *BCR/ABL* fusion gene could sometimes be important in determining disease progression.

In summary, blast crisis CML appears to be associated with some degree of alteration in *BCR/ABL* gene expression which can arise through duplication of the Ph chromosome, changes in splicing of mRNA transcripts, and deregulation of gene expression. Whether or not these are critical in determining a more aggressive cell proliferation or whether other cytogenetic changes, oncogene activations or other changes are of greater importance remains to be determined.

IX. EXPERIMENTAL STUDIES WITH *BCR/ABL*

Cloned cDNA derived from *BCR/ABL* fusions can be used in experimental studies
to assess the effects of the introduction of the fusion gene in various cell types.
Plasmid constructs are made where *BCR/ABL* cDNA is joined to appropriate
replication and promoter sequences together with genes encoding for antibiotic
resistance to assist in selection of transfected cells.

When NIH 3T3 fibroblasts are transfected with a construct that encodes P210,
expression of P210 can be demonstrated, but the cells show no morphological signs
of transformation.[180] However, when a *gag* determinant is fused to *BCR/ABL*,
transformation of NIH 3T3 cells is achieved with cells showing altered morphology
similar to that seen when cells are infected with A-MuLV.[180] These results indicate
that myristylation-dependent membrane localization is probably necessary for
BCR/ABL transformation of fibroblasts.

In contrast to the experiments with NIH 3T3 cells, a *gag* determinant does not to
appear to be necessary for the transformation of bone marrow cells. Transformation
characteristics are seen when fresh bone marrow cells are infected and maintained
under the conditions used for long-term culture of B-lymphoid cell lines.[181] Variability is observed in the ability to grow in soft agar and in the induction of
lymphomas after inoculation of the transformed cells into syngeneic mice. Transformation is also achieved with long-term B-cell cultures and long-term myeloid
cell cultures although the type of transformed cell appears to be an early B cell for
either type of culture.[182] A murine bone marrow derived cell line thought to be of
early B-cell lineage referred to as Ba/F3 can also be transformed by *BCR/ABL*
retroviral constructs.[183] This cell line normally has a requirement for interleukin-3
for growth but after transformation becomes independent of the growth factor. This
observation suggests that the P210 protein provides an alternative to interleukin-3
for growth stimulation. The transformed Ba/F3 cells induced tumors in nude mice
after a short latency.

Differences appear to exist between the P210 and P190 proteins in terms of their
potency to transform cells.[184] Using retrovirus vectors with structures that differed
only in the size of the *BCR* gene component, McLaughlin et al.[184] compared the
outgrowth of cells transformed by the two types of construct and showed that the
growth stimulating effect of P190 was significantly greater than that for P210. P190
appeared to be more effective than P210 in the induction of tumors following
inoculation into mice, but, as in the case of P210, the presence of P190 in transfected
cells did not invariably result in tumors. The stronger stimulating effects of P190
compared to P210 would be consistent with the more aggressive nature of the
leukemias associated with P190 (ALL and AML) compared to most of the
leukemias associated with P210 (CML) at presentation. The fact that tumors are
not invariably present in mice carrying the transformed cells would seem to indicate
a need for additional oncogene activations for tumor formation.[184]

In the *in vitro* studies described above, the cell phenotype of transformed bone marrow cells was lymphoid in nature with the induction of lymphomas when cells were reintroduced to mice. In transgenic mice containing *BCR/v-abl* constructs, T-cell lymphomas and some pre-B-cell lymphomas were found.[185] An alternative model system has been developed which does not limit the tumor type to lymphomas.[186–189] When mouse bone marrow is treated with 5-fluorouracil, transfected, and then returned to irradiated recipients, a diverse set of transformed hematologic cell types is seen. Elefanty et al.[186] noted macrophage, erythroid, mast cell, pre-B lymphoid, T-lymphoid, and mixed lineage tumors in their experiments. They noted that the spectrum of tumors observed differed substantially in different stains of mice. Using BALB/c mice, Daley et al.[187] have noted the induction of three disease types: (1) a myeloproliferative syndrome with a mean latency of approximately 9 weeks, (2) an acute lymphoblastic leukemia arising after a mean of 14 weeks, and (3) macrophage derived tumors after an average of 16.5 weeks. Similar results were reported by Kelliher et al.[188] The myeloproliferative disorder strongly resembles chronic phase CML with the animals showing enlarged spleen, increased granulocytes in peripheral blood, and hypercellular bone marrow containing myeloid cells at all stages of differentiation.

If the CML-like disorder is transplanted to irradiated syngeneic recipients, a small proportion of animals continue to demonstrate the chronic CML-like disorder or may show an acute leukemia of lymphoid or myeloid type—i.e., the disorder appears to evolve to resemble blast crisis CML.[189] The development of the animal model for CML will clearly be helpful not only in further understanding the role of *BCR/ABL* fusions but also in providing a model for testing various therapeutic strategies.

Selective inhibition of *BCR/ABL* fusion gene under *in vitro* conditions provides additional evidence that expression of *BCR/ABL* is directly associated with the leukemic cell phenotype. Szczylik et al.[190] prepared oligodeoxynucleotides complimentary to mRNA (specifically, the sequences corresponding to the *BCR/ABL* junctions) and tested the effect of these antisense oligomers on the outgrowth of *in vitro* colonies from CML blast crisis patients. Fewer colonies with reduced numbers of cells were seen in cultures where a matched antisense oligomer was present compared to control cultures. Residual colonies in the treated cultures lacked detectable *BCR/ABL* mRNA. Studies of this type clearly have important therapeutic implications; gene-targeted anti-leukemic therapy appears to be possible.

X. OVERVIEW

Studies on colon carcinomas and other tumor types has led to a model for cancer in which there is an accumulation of critical mutations with gradual progression to full expression of the cancer cell phenotype.[191] Often, the order in which the mutations arise is unimportant.[192] The *BCR/ABL* fusion that characterizes many human leukemias differs in that this alteration is usually present at diagnosis with subsequent additional mutations occurring during evolution. A few cases of CML preleukemia and acute leukemia have been described that appear to be Ph-nega-tive[193] at diagnosis and subsequently acquired a Ph chromosome. The rarity of these cases attests to the role of *BCR/ABL* as a very early event in the disease pathogenesis in the vast majority of cases.

Whether or not *BCR/ABL* fusion is truly the seminal change cannot be easily proven. Fialkow et al.[194] noted in a Ph-positive CML patient, a high proportion of Ph-negative B-lymphoid clones showing a particular glucose-6-phosphate dehy-drogenase (G6PD) isozyme, identical in type to that seen in the Ph-positive clone. From this observation they suggested that a primary alteration causing proliferation of a pluripotent stem cell proceeds the acquisition of a Ph chromosome. However, distortion in G6PD isozyme types within any particular tissue may not necessarily indicate a neoplastic change that enhanced the proliferative capacity of a stem cell clone. Currently, there remains little data to support the concept of specific altera-tions preceding the *BCR/ABL* gene fusion.

It is also apparent that retention of the *BCR/ABL* fusion gene is the usual situation. A few patients have been described in which there may be a loss of the Ph chromosome.[157–159] Given the fact that a *BCR/ABL* fusion is not essential for acute leukemia, and that there must be alternative pathways to develop the leukemic cell phenotype, it is not unexpected that during clonal evolution additional changes could occur within a Ph-positive population that provides an equal or greater opportunity for uncontrolled cell proliferation. Within such an evolving cell clone, the selective advantage in retaining the Ph chromosome may no longer exist.

The fact that a single oncogene activation usually characterizes the earliest stages of a disease and that the change usually persists within the clone may eventually prove to be advantageous. While traditional therapeutic strategies have concen-trated on abnormal cell eradication, new approaches such as the use of antisense oligomers raise the exciting possibility of selectively inhibiting the production or activity of common aberrant oncoproteins.[190]

Clearly, advances in therapeutics can only be made with a thorough under-standing of the details of the abnormalities in the cells. The advancements made in the understanding the role of the *BCR/ABL* oncogene have provided a paradigm for cancer research.

REFERENCES

1. Silver, R. T. Chronic myeloid leukemia. A Perspective of the clinical and biological issues of the chronic phase. *Hematol./Oncol. Clinics of N.A.* **1990**, *4*, 319–335.
2. Silver, R. T.; Gale, R. P. Chronic myeloid leukemia. *Am. J. Med.* **1986**, 80, 1137–1148.
3. Silver, R. T. Chronic myeloid leukemia. In: *Contemporary Issues in Clinical Oncology* (Wiernick, Ed.). Churchill Livingston, New York, 1985, p. 227.

4. Galbraith, P. R.; Abu Zahra, H. T. Granulocytopoiesis in chronic granulocytic leukemia. *Br. J. Hematol.* **1972**, *22*, 135–143.

5. Li, F. P. The chronic leukemias: etiology and epidemiology. In: *Neoplastic Diseases of the Blood* (Wiernik, P.H.; Canellos, G.P.; Kyle, R.A.; Schiffer, C.A., Eds.). Churchill Livingston, New York, 1985, Vol. 1, pp. 7–17.

6. Champlin, R. E.; Golde, D. W. Chronic myelogenous leukemia: recent advances. *Blood* **1985**, *65*, 1039–1047.

7. Sokal, J. E.; Baccarani, M.; Russo, D.; Tura, S. Staging in chronic myelogenous leukemia. *Semin. Hemat.* **1988**, *25*, 49–61.

8. Karanas, A.; Silver, R. T. Characteristics of the terminal phase of chronic granulocytic leukemia. *Blood* **1968**, *32*, 445–459.

9. Gralnick, H. R.; Bennett, J. M. Bone marrow histology in chronic granulocytic leukemia: observations on myelofibrosis and the accelerated phase. *AEC Symposium Series* **1970**, *19*, 583–598.

10. Arlin, Z. A.; Silver, R. T.; Bennett, J. M. Blastic phase of chronic myeloid leukemia (bl CML): a proposal for standardization of diagnostic and response criteria. *Leukemia* **1990**, *4*, 755–757.

11. Canellos, G. P. Chronic granulocytic leukemia. *Med. Clin. N.A.* **1976**, *60*, 1001.

12. Greaves, M. F. "Target" cells, differentiation and clonal evolution in chronic granulocytic leukemia: a "model" for understanding the biology of malignancy. p15 In: *Chronic Granulocytic Leukemia* (Shaw, M. T., Ed.). Praeger Publishing, East Sussex, UK 1982.

13. Janossy, G.; Woodruff, R. K.; Paxton, A.; et al. Membrane marker and cell separation studies in Ph[1] positive leukemia. *Blood* **1978**, *51*, 861–877.

14. Griffin, J. D.; Todd, R. F.; Ritz, J.; et al. Differentiation patterns in the blastic phase of chronic myeloid leukemia. *Blood* **1983**, *61*, 85–91.

15. Parreira, L.; Kearney, L.; Rassool, F.; et al. Correlation between chromosomal abnormalities and blast phenotype in the blast crisis of Ph-positive CGL. *Cancer Genet. Cytogenet.* **1986**, *22*, 29–34.

16. Bernstein, R. Cytogenetics of chronic myelogenous leukemia. *Semin. Hematol.* **1988**, *25*, 20–34.

17. Third International Workshop on Chromosomes in Leukemia, Lund Sweden, July 21–25 1981. Clinical significance of chromosome abnormalities in acute lymphoblastic leukemia. *Cancer Genet. Cytogenet.* **1981**, *4*, 111–137.

18. Rowley, J. D. Biological implications of consistent chromosome rearrangements in leukemia and lymphoma. *Cancer Res.* **1984**, *44*, 3159–3168.

19. Lowagie, A.; Criel, A.; Verfaillie, C. M.; et al. Philadelphia-positive T-acute lymphoblastic leukemia. *Cancer Genet. Cytogenet.* **1985**, *16*, 297–300.

20. Bloomfield, C. D.; Brunning, R. D.; Smith, K. A.; Nesbit, M. E. Prognostic significance of the Philadelphia chromosome in acute lymphocytic leukemia. *Cancer Genet. Cytogenet.* **1980**, *1*, 229–238.

21. Helenglass, G.; Testa, J. R.; Schiffer, C. A. Philadelphia chromosome positive acute leukemia: morphologic and clinical correlations. *Am. J. Hematol.* **1987**, *25*, 311–324.

22. Catovsky, D. Ph[1]-positive acute leukemia and chronic granulocytic leukemia: One or two diseases? *Br. J. Haematol.* **1979**, *42*, 493–498.

23. Ribeiro, R. C.; Abromowitch, M.; Raimondi, S. C.; Murphy, S. B.; Behm, F.; Williams, D. L. Clinical and biologic hallmarks of the Philadelphia chromosome in childhood acute lymphoblastic leukemia. *Blood* **1987**, *70*, 948–953.

24. Fourth International Workshop on Chromosomes in Leukemia (1982). A prospective study of acute nonlymphocytic leukemia. *Cancer Genet. Cytogenet.* **1984**, *11*, 249–360.

25. Sasaki, M.; Kondo, K.; Tomiyasu, T. Cytogenetic characterization of ten cases of Ph[1]-positive acute myelogenous leukemia. *Cancer Genet. Cytogenet.* **1983**, *9*, 119–128.

26. Heath, C. W. Epidemiology and hereditary aspects of acute leukemia. In: *Neoplastic Diseases of the Blood* (Wiernik, P.H.; Canellos, G.P.; Kyle, R.A.; Schiffer, C.A., Eds.). Churchill Livingston, New York, 1985, Vol. 1.

27. Nowell, P. C.; Hungerford, D. A. A minute chromosome in human chronic granulocytic leukemia. *Science* **1960**, *132*, 1497.

28. Rowley, J. D. A new consistent chromosomal abnormality in chronic myelogenous leukaemia identified by quinacrine fluorescence and Giemsa staining. *Nature* **1973**, *243*, 290–291.

29. Prakash, O.; Yunis, J. J. High resolution chromosomes of the t(9;22) positive leukemias. *Cancer Genet. Cytogenet.* **1984**, *11*, 361–367.

30. De Braekeleer, M. Breakpoint distribution in variant Philadelphia translocations in chronic myeloid leukemia. *Cancer Genet. Cytogenet.* **1986**, *23*, 167–170.

31. Verma, R. S.; Macera, M. J. Genomic diversity of Philadelphia-positive chronic myelogenous leukemia. *Leuk. Res.* **1987**, *11*, 833–842.

32. Hayata, I.; Kakati, S.; Sandberg, A. A. A new translocation related to the Philadelphia chromosome. *Lancet* **1973**, *2*, 1385.

33. Bernstein, R.; Pinto, M. R.; Rosendorff, J.; et al. "Masked" Ph[1] chromosome abnormalities in CML: A report of two unique cases. *Blood* **1984**, *63*, 399–406.

34. Dreazen, O.; Klisak, I.; Rassool, F.; et al. Do oncogenes determine clinical features in chronic myeloid leukaemia? *Lancet* **1987**, *1*, 1402–1405.

35. Morris, C. M.; Reeve, A. E.; Fitzgerald, P. H.; et al. Genomic diversity correlates with clinical variation in Ph[1] negative chronic myeloid leukaemia. *Nature* **1986**, *320*, 281–283.

36. Weinstein, M. E.; Grossman, A.; Perle, M. A.; et al. The karyotype of Philadelphia chromosome-negative, bcr rearrangement-positive chronic myeloid leukemia. *Cancer Genet. Cytogenet.* **1988**, *35*, 223–229.

37. Morris, C. M.; Heisterkamp, N.; Kennedy, M. A.; Fitzgerald, P. H.; Griffin, J. Ph-negative chronic myeloid leukemia: molecular analysis of ABL insertion into M-BCR on chromosome 22. *Blood* **1990**, *76*, 1812–1818.

38. Lubbert, M.; Miller, C. W.; Crawford, L.; Koeffler, H. P. p53 in chronic myelogenous leukemia. Study of mechanisms of differential expression. *J. Exp. Med.* **1988**, *167*, 873–886.

39. Kilman, Z.; Prokocimer, M.; Peller, S.; et al. Rearrangement in the p53 gene in Philadelphia chromosome positive chronic myelogenous leukemia. *Blood* **1989**, *74*, 2318–2324.

40. Liu, E.; Hjelle, B.; Bishop, J. M. Transforming genes in chronic myelogenous leukemia. *Proc. Natl. Acad. Sci. USA* **1988**, *85*, 1952–1956.

41. Cogswell, P. C.; Morgan, R.; Dunn, M.; et al. Mutations of the *Ras* protooncogenes in chronic myelogenous leukemia: a high frequency of *Ras* mutations in *bcr/abl* rearrangement-negative chronic myelogenous leukemia. *Blood* **1989**, *74*, 2629–2633.

42. Sandberg, A. A.; Morgan, R.; Kipps, T. J.; Hecht, B. K.; Hecht, F. The Philadelphia (Ph) chromosome in leukemia II. Variant Ph translocations in acute lymphoblastic leukemia. *Cancer Genet. Cytogenet.* **1985**, *14*, 11–21.

43. Rowley, J. D. Ph[1]-positive leukaemia, including chronic myelogenous leukemia. *Clin. Haematol.* **1980**, *9*, 55–86.

44. Abelson, H. T.; Rabstein, L. S. Lymphosarcoma: virus induced thymic dependent disease in mice. *Cancer Res.* **1970**, *30*, 2213–2222.

45. Whitlock, C. A.; Witte, O. N. The complexity of virus-cell interactions in Abelson virus infection of lymphoid and other hematopoietic cells. *Adv. Immunol.* **1985**, *37*, 73–98.

46. Goff, S. The Abelson murine leukemia virus oncogene. *Proc. Soc. Exp. Biol. Med.* **1985**, *179*, 403–412.

47. Rosenberg, N.; Witte, O. N. The viral and cellular forms of the Abelson (*abl*) oncogene. *Adv. Virus Res.* **1988**, *35*, 39–81.

48. Ramakrishnan, L.; Rosenberg, N. *abl* genes. *Biochimica et Biophysica Acta* **1989**, *989*, 209–224.

49. Scher, C. D.; Siegler, R. S. Direct formation of 3TS cells by Abelson murine leukemia virus. *Nature* **1975**, *252*, 729–731.

50. Rosenberg, N.; Baltimore, D.; Scher, C. D. *In vitro* transformation of lymphoid cells by Abelson murine leukemia virus. *Proc. Natl. Acad. Sci. USA* **1975**, *72*, 1932–1936.

51. Rosenberg, N.; Baltimore, D. A quantitative assay for transformation or bone marrow cells by Abelson murine leukemia virus. *Exp. Med.* **1976**, *143*, 1453–1463.

52. Witte, O. N.; Degupta, A.; Baltimore, D. Abelson murine leukemia virus protein is phosphorylated *in vitro* to form phosphotyrosine. *Nature* **1980**, *283*, 826–831.

53. Van De Ven, W. J. M.; Reynolds, F. R.; Stephenson, J. R. The nonstructural components of polyproteins encoded by replication-defective mammalian transforming retroviruses are phosphorylated and have associated protein kinase activity. *Virology* **1980**, *101*, 185–197.

54. Hanks, S. K.; Quinn, A. M.; Hunter, T. The protein kinase family: conserved features and delayed phylogeny of the catalytic domains. *Science* **1988**, *241*, 42–51.

55. Wang, J. Y. J.; Ledley, F.; Goff, S.; Lee, R.; Groner, Y.; Baltimore, D. The mouse c-abl locus: molecular cloning and characterization. *Cell* **1984**, *36*, 349–356.

56. Reddy, E. P.; Smith, M. J.; Srinivasan, A. Nucleotide sequence of Abelson murine leukemia virus genome: structural similarity of its transforming gene product to other *onc* gene products with tyrosine-specific kinase activity. *Proc. Natl. Acad. Sci. USA* **1983**, *80*, 3623–3627.

57. Jackson, P.; Baltimore, D. N-terminal mutations activate the leukemia potential of the myristlated form of c-abl. *EMBO J.* **1989**, *8*, 449–456.

58. Besmer, P.; Hardy, W. D.; Zuckerman, E. E.; Bergold, P.; Lederman, L.; Snyder, H. W. The Hardy-Zuckerman 2-FesV, a new feline retrovirus with oncogene homology to Abelson-MuLV. *Nature* **1983**, *303*, 825–828.

59. Bergold, P. J.; Blumenthal, J. A.; D'Andrea, E.; et al. Nucleic acid sequence and oncogenic properties of the HZ2 feline sarcoma virus v-*abl* insert. *J. Virol.* **1987**, *61*, 1193–1202.

60. Ponticelli, A. S.; Whitlock, C. A.; Rosenberg, N.; Witte, O. *In vivo* tyrosine phosphorylations of the Abelson virus transforming protein are absent in its normal cellular homolog. *Cell* **1982**, *29*, 953–960.

61. Konopka, J. B.; Witte, O. N. Activation of the *abl* oncogene in murine and human leukemias. *Biochimica Biophysica Acta* **1985**, *828*, 1–17.

62. Bernards, A.; Rubin, C. M.; Westbrook, C. A.; Paskind, M.; Baltimore, D. The first intron in the human c-abl gene is at least 200 kilobases long and is a target for translocations in chronic myelogenous leukemia. *Mol. Cell. Biol.* **1987**, *7*, 3231–3236.

63. Heisterkamp, N.; Stephenson, J. R.; Groffen, J.; et al. Localization of the c-abl oncogene adjacent to a translocation break point in chronic myelocytic leukaemia. *Nature* **1983**, *306*, 239–242.

64. Shtivelman, E.; Lifshitz, B.; Gale, R. P.; Roe, B. A.; Canaani, E. Alternative splicing of RNAs transcribed from the human abl gene and from the *bcr-abl* fused gene. *Cell* **1986**, *47*, 277–284.

65. Muller, R.; Slamon, D. J.; Tremblay, J. M.; Cline, M. J.; Verma, I. M. Differential expression of cellular oncogenes during pre- and postnatal development of the mouse. *Nature* **1982**, *299*, 640–644.

66. Wang, J. Y. J.; Baltimore, D. Cellular RNA homologous to the Abelson murine leukemia virus transforming gene: expression and relationship to the viral sequence. *Mol. Cell. Biol.* **1983**, *3*, 773–779.

67. Renshaw, M. W.; Capozza, M. A.; Wang, J. Y. J. Differential expression of type specific c-*abl* mRNA's in mouse tissues and cell lines. *Mol. Cell. Biol.* **1988**, *8*, 4547–4551.

68. Konopka, J. B.; Witte, O. N. Detection of c-abl tyrosine kinase activity in vitro permits direct comparison of normal and altered abl gene products. *Mol. Cell. Biol.* **1985**, *5*, 3116–3123.

69. Rees-Jones, R. W.; Goff, S. P. Insertional mutagenesis of the Abelson murine leukemia virus genome: identification of mutants with altered kinase activity and defective transforming ability. *J. Virol.* **1988**, *62*, 978–986.

70. Franz, W. M.; Berger, P.; Wang, J. W. D. Deletions of an N-terminal regulatory domain of the c-*abl* tyrosine kinase activates its oncogenic potential. *EMBO J.* **1989**, *8*, 137–147.

71. Kruh, G. D.; King, C. R.; Kraus, M. H.; et al. A novel human gene closely related to the *abl* proto-oncogene. *Science* **1986**, *234*, 1545–1548.

72. Mitelman, F.; Kaneko, Y.; Trent, J. M. Report of the committee on chromosome changes in neoplasia. Human Gene Mapping 11. *Cytogenet. Cell. Genet.* **1990**, *55*, 358–386.

73. Hsu, L. Y.; Pinchiaroli, D.; Gilbert, H. S.; Wittman, R.; Hirschhorn, K. Partial trisomy of the long arm of chromosome 1 in myelofibrosis and polycythemia vera. *Am. J. Hematol.* **1977**, *2*, 375–383.

74. Gahrton, G.; Friberg, K.; Lindsten, J.; Zech, L. Duplication of part of the long arm of chromosome 1 in myelofibrosis terminating in acute myeloblastic leukemia. *Hereditas* **1978**, *88*, 1–5.

75. Heisterkamp, N.; Stam, K.; Groffen, J.; de Klein, A.; Grosveld, G. Structural organization of the bcr gene and its role in the Ph[1] translocation. *Nature* **1985**, *315*, 758–761.

76. Groffen, J.; Stephenson, J. R.; Heisterkamp, N.; de Klein, A.; Bartram, C. R.; Grosveld, G. Philadelphia chromosomal breakpoints are clustered within a limited region, bcr, on chromosome 22. *Cell* **1984**, *36*, 93–99.

77. Hariharan, I. K.; Adams, J. M. cDNA sequence for human bcr, the gene that translocates to the abl oncogene in chronic myeloid leukemia. *EMBO J.* **1987**, *6*, 115–119.

78. Mes-Masson, A. M.; McLaughlin, J.; Daley, G. Q.; Paskind, M.; Witte, O. N. Overlapping cDNA clones define the complete coding region for the $P210^{c-abl}$ gene product associated with chronic myelogenous leukemia cells containing the Philadelphia chromosome. *Proc. Natl. Acad. Sci. USA* **1986**, *83*, 9768–9772.

79. Heisterkamp, N.; Knoppel, E.; Groffen, J. The first BCR gene intron contains breakpoints in Philadelphia chromosome positive leukemia. *Nucleic Acids Res.* **1988**, *16*, 10069–10081.

80. Collins, S.; Coleman, H.; Groudine, M. Expression of bcr and bcr-abl fusion transcripts in normal and leukemic cells. *Mol. Cell. Biol.* **1987**, *7*, 2870–2876.

81. Shtivelman, E.; Lifshitz, B.; Gale, R. P.; Canaani, E. Fused transcript of abl and bcr genes in chronic myelogenous leukemia. *Nature* **1985**, *315*, 550–554.

82. Benn-Neriah, Y.; Daley, G. Q.; Mes-Masson. A. M.; Witte, O. N.; Baltimore, D. The chronic myelogenous leukemia-specific p210 protein is the product of the bcr/abl hybrid gene. *Science* **1986**, *233*, 212–214.

83. Stam, K.; Heisterkamp, N.; Reynolds, F. H., Jr.; Groffen, J. Evidence that the Ph[1] gene encodes a 160,000-dalton phosphoprotein with associated kinase activity. *Mol. Cell. Biol.* **1987**, *7*, 1955–1960.

84. Diekmann, D.; Brill, S.; Garrett, M.; et al. Bcr encodes a GTPase-activity protein for $p21^{rac}$. *Nature* **1991**, *351*, 400–402.

85. Hall, A. The cellular functions of small GTP-binding proteins. *Science* **1990**, *249*, 635–640.

86. Barbacid, M. ras genes. *Ann. Rev. Biochem.* **1987**, *56*, 779–782.

87. Didsbury, J.; Weber, R. F.; Bokoch, G. M.; Evans, T.; Snyderman, R. rac, a novel ras-related family of proteins that are botulinum toxin substrates. *J. Biol. Chem.* **1989**, *264*, 16378–16382.

88. Croce, C. M.; Huebner, K.; Isobe, M.; et al. Mapping of four BCR-related loci to chromosome region 22q11: order of BCR loci relative to chronic myelogenous leukemia and acute lymphoblastic leukemia breakpoints. *Proc. Natl. Acad. Sci. USA* **1987**, *84*, 7174–7178.

89. Heisterkamp, N.; Groffen, J.; Stephenson, J. R.; et al. Chromosomal localization of human cellular homologues of two viral oncogenes. *Nature* **1982**, *299*, 747–749.

90. de Klein, A.; Geurts van Kessel, A.; Grosveld, G.; et al. A cellular oncogene is translocated to the Philadelphia chromosome in chronic myelocytic leukaemia. *Nature* **1982**, *300*, 765–767.

91. Bartram, C. R.; de Klein, A.; Hagemeijer, A.; et al. Translocation of c-abl oncogene correlates with the presence of a Philadelphia chromosome in chronic myelogenous leukemia. *Nature* **1983**, *306*, 277–280.

92. Gale, R. P.; Canaani, E. An 8-kilobase abl RNA transcript in chronic myelogenous leukemia. *Proc. Natl. Acad. Sci. USA* **1984**, *81*, 5648–5652.

93. Canaani, E.; Gale, R. P.; Steiner-Saltz, D.; Berrebi, A.; Aghai, E.; Januszewicz, E. Altered transcription of an oncogene in chronic myeloid leukemia. *Lancet* **1984**, *1*, 593–595.

94. Collins, S. J.; Kubonishi, I.; Miyoshi, I.; Groudine, M. T. Altered transcription of the c-abl oncogene in K-562 and other chronic myelogenous leukemia cells. *Science* **1984**, *225*, 72–74.

95. Stam, K.; Heisterkamp, N.; Grosveld, G.; et al. Evidence of a new chimeric *bcr/c-abl* mRNA in patients with chronic myelocytic leukemia and the Philadelphia chromosome. *N. Engl. J. Med.* **1985**, *313*, 1429–1430.

96. Konopka, J. B.; Watanabe, S. M.; Witte, O. N. An alteration of the human c-abl protein in K562 leukemia cells unmasks associated tyrosine kinase activity. *Cell* **1984**, *37*, 1035–1042.

97. Kloetzer, W.; Kurzrock, R.; Smith, L.; et al. The human cellular abl gene product in the chronic myelogenous leukemia cell line K562 has an associated tyrosine protein kinase activity. *Virology* **1985**, *140*, 230–238.

98. Kurzrock, R.; Gutterman, J. U.; Talpaz, M. The molecular genetics of Philadelphia chromosome-positive leukemias. *New. Eng. J. Med.* **1988**, *319*, 990–998.

99. Schaefer-Rego, K.; Dudek, H.; Popenoe, D.; et al. CML patients in blast crisis have breakpoints localized to a specific region of the bcr. *Blood* **1987**, *70*, 448–455.

100. Benn, P.; Soper, L.; Eisenberg, A.; et al. Utility of molecular genetic analysis of bcr arrangement in the diagnosis of chronic myeloid leukemia. *Cancer Genet. Cytogenet.* **1987**, *29*, 1–7.

101. Dobrovic, A.; Trainor, K. J.; Morley, A. A. Detection of the molecular abnormality in chronic myelogenous leukemia by use of the polymerase chain reaction. *Blood* **1988**, *72*, 2063–2065.

102. Kawasaki, E. S.; Clark, S. S.; Coyne, M. Y.; et al. Diagnosis of chronic myeloid and acute lymphocytic leukemias by detection of leukemia-specific mRNA sequences amplified in vitro. *Proc. Natl. Acad. Sci. USA* **1988**, *85*, 5698–5702.

103. Hermans. A.; Selleri, L.; Gow, J.; et al. Absence of alternative splicing in bcr-abl mRNA in chronic myeloid leukemia cell lines. *Blood* **1988**, *72*, 2066–2069.

104. Hooberman, A. L.; Carrino, J. J.; Leibowitz, D.; et al. Unexpected heterogeneity of BCR-ABL fusion mRNA detected by polymerase chain reaction in Philadelphia chromosome positive acute lymphoblastic leukemia. *Proc. Natl. Acad. Sci. USA* **1989**, *86*, 4259–4263.

105. Shtivelman, E.; Lifshitz, B.; Gale, R.; et al. Alternative splicing of RNAs transcribed from the human abl gene and from the bcr-abl fused gene. *Cell* **1986**, *47*, 277–284.

106. Bartram, C. R.; Bross-Bach, U.; Schmidt, H.; Waller, H. D. Philadelphia-positive chronic myelogenous leukemia with breakpoint 5′ of the breakpoint cluster region but within the *bcr* gene. *Blut* **1987**, *55*, 505–511.

107. Shtalrid, M.; Talpaz, M.; Kurzrock, R.; et al. Analysis of breakpoints within the bcr gene and correlation with clinical course in Ph-positive chronic myelogenous leukemia. *Blood* **1988**, *72*, 485–490.

108. Sagio, G.; Guerrasio, A.; Tassinari, A.; et al. Variability of the molecular defects corresponding to the presence of a Philadelphia chromosome in human hematologic malignancies. *Blood* **1988**, *72*, 1203–1208.

109. Leibowitz, D. Molecular diagnosis of chronic myelocytic leukemia (CML). In: *Molecular Genetics in Cancer Diagnosis* (Cossman, J., Ed.). Elsevier, New York, 1990, pp. 179–188.

110. Eisenberg, A.; Silver, R.; Soper, L.; et al. The location of breakpoints within the breakpoint cluster region (bcr) of chromosome 22 in chronic myeloid leukemia. *Leukemia* **1988**, *2*, 642–647.

111. Mills, K. I.; MacKenzie, E. D.; Birnie, G. D. The site of the breakpoint within the bcr is a prognostic factor in Philadelphia-positive CML patients. *Blood* **1988**, *72*, 1237–1241.

112. Selleri, L.; Narni, F.; Emilia, G.; et al. Philadelphia-positive chronic myeloid leukemia with a chromosome 22 breakpoint outside the breakpoint cluster region. *Blood* **1987**, *70*, 1659–1664.

113. Mills, K. I.; Hynds, S. A.; Burnett, A. K.; MacKenzie, E. D.; Birnie, G. D. Further evidence that the site of the breakpoint in the major breakpoint cluster region (M-bcr) may be a prognostic factor. *Leukemia* **1989**, *3*, 837–840.

114. Jiang, X.; Trujillo, J. M.; Liang, J. C. Chromosome breakpoint within the first intron of the ABL gene are nonrandom in patients with chronic myelogenous leukemia. *Blood* **1990**, *76*, 597–601.

115. Van der Plas, D. C.; Soekarman, D.; van Gent, A. M.; Grosveld, G.; Hagemeijer, A. *bcr-abl* mRNA lacking *abl* exon A2 detected by polymerase chain reaction in a chronic myelogenous leukemia patient. *Leukemia* **1991**, *5*, 457–461.

116. Popencoe, D. W.; Schaefer-Rego, K.; Mears, J. G.; Bank, A.; Leibowitz, D. Frequent and extensive deletion during the 9,22 translocation in CML. *Blood* **1986**, *68*, 1123–1128.

117. Grossman, A.; Silver, R. T.; Arlin, Z.; et al. Fine mapping of chromosome 22 breakpoints within the breakpoint cluster region (bcr) implies a role for exon 3 in determining disease duration in chronic myeloid leukemia. *Am. J. Hum. Genet.* **1989**, *45*, 729–738.

118. Mills, K. I.; Sproul, A. M.; Leibowitz, D.; Burnett, A. K. Mapping of breakpoints, and relationship to BCR-ABL RNA expression in Philadelphia-chromosome positive chronic myeloid leukemia patients with a breakpoint around exon 14 (b3) of the BCR gene. *Leukemia* **1991**, *5*, 937–941.

119. Chen, S. J.; Chen, Z.; d'Auriol, L.; Le Coniat, M.; Grausz, D.; Berger, R. Ph1[+] bcr[−] acute leukemias: Implication of Alu sequences in a chromosomal translocation occurring in the new cluster region within the BCR gene. *Oncogene* **1989**, *4*, 195–202.

120. van der Feltz, M. J. M.; Shivji, M. K. K.; Allen, P. B.; Heisterkamp, N.; Groffen, J.; Wiedemann, L.M. Nucleotide sequence of both reciprocal translocation junction regions in a patient with Ph positive acute lymphoblastic leukemia, with a breakpoint within the first intron of the BCR gene. *Nucleic Acids Res.* **1989**, *17*, 1–10.

121. Papadopoulos, P. C.; Greenstein, A. M.; Gaffney, R. A.; Westbrook, C. A.; Weidemann, L. M. Characterization of the translocation breakpoint sequences in Philadelphia-positive acute lymphoblastic leukemia. *Genes, Chromosomes and Cancer* **1990**, *1*, 233–239.

122. Rabbitts, T. H.; Boehm, T.; Mengle-Gaw, L. Chromosome abnormalities in lymphoid tumors: mechanism and role in tumor pathogenesis. *Trends in Genetics* **1988**, *4*, 300–304.

123. Shtivelman, E.; Gale, R. P.; Dreazen, O.; et al. *bcr-abl* RNA in patients with chronic myelogenous leukemia. *Blood* **1987**, *69*, 971–973.

124. Lange, W.; Snyder, D. S.; Castro, R.; Rossi, J. J.; Blume, K. G. Detection by enzyme amplification of *bcr-abl* mRNA in peripheral blood and bone marrow cells from patients with chronic myelogenous leukemia. *Blood* **1989**, *73*, 1735–1741.

125. Romero, P.; Beren, M.; Shtalrid, M.; Andersson, B.; Talpez, M.; Blick, M. Alternative 5′ end of the *bcr-abl* transcript in chronic myelogenous leukemia. *Oncogene* **1989**, *4*, 93–98.

126. Hariharan, I.; Adams, J. M. cDNA sequence for human *bcr*, the gene that translocates to the *abl* oncogene in chronic myeloid leukemia. *EMBO J.* **1987**, *6*, 115–119.

127. Konopka, J. B.; Witte, O. N. Detection of c-abl tyrosine kinase activity in vitro permits direct comparison of normal and altered abl gene products. *Mol. Cell. Biol.* **1985**, *5*, 3116–3123.

128. Maxwell, S. A.; Kurzrock, R.; Parsons, S. J.; et al. Analysis of P210bcr-abl tyrosine protein kinase activity in various subtypes of Philadelphia chromosome-positive cells from chronic myelogenous leukemia patients. *Cancer Res.* **1987**, *47*, 1731–1739.

129. Van Denderen, J.; Hermans, A.; Meeuwsen, T.; et al. Antibody recognition of the tumor-specific bcr-abl joining region in chronic myeloid leukemia. *J. Exp. Med.* **1989**, *169*, 87–98.

130. Van Denderen, J.; Van der Plas, D.; Meeuwsen, T.; et al. Immunologic characterization of the tumor-specific *bcr-abl* junction in Philadelphia chromosome-positive acute lymphoblastic leukemia. *Blood* **1990**, *76*, 136–141.

131. Blennerhassett, G. T.; Furth, M. E.; Anderson, A. et al. Clinical evaluation of a DNA probe assay for the Philadelphia (Ph[1]) translocation in chronic myelogenous leukemia. *Leukemia* **1988**, *2*, 648–657.

132. Wiedemann, L. M.; Karhi, K. K.; Shivji, M. K. K.; et al. The correlation of breakpoint cluster region rearrangement and p210 *phl/abl* expression with morphological analysis of Ph-negative chronic myeloid leukemia and other myeloproliferative diseases. *Blood* **1988**, *71*, 349–355.

133. Sandberg, A.A. Chromosomes and causation of human cancer and leukemia: XL. The Ph[1] and other translocations in CML. *Cancer* **1980**, *46*, 2221–2226.

134. Kurzrock, R.; Kantarjian, H. M.; Shtalrid, M.; Gutterman, J. U.; Talpaz, M. Philadelphia chromosome-negative chronic myelogenous leukemia without breakpoint cluster region rearrangement: a chronic myeloid leukemia with a distinct clinical course. *Blood* **1990**, *75*, 445–452.

135. Rowley, J. D. Recurring chromosome abnormalities in leukemia and lymphoma. *Semin. Hematol.* **1990**, *27*, 122–136.

136. Roschmann, E.; Assum, G.; Fink, T. RFLP detected with a 5'-bcr-gene-sequence (HGM8 provisional no. D22S11). *Nucleic Acids Res.* **1987**, *15*, 1883.

137. Kato, Y.; Sawada, H.; Tashima, M.; et al. Restriction fragment length polymorphism of bcr in Japanese patients with hematological malignancies. *Leukemia* **1988**, *2*, 701–703.

138. Benn, P.; Grossman, A.; Soper, L.; Halka, K.; Eisenberg, A.; Gascon, P. A rare restriction enzyme site polymorphism in the breakpoint cluster region (bcr) of chromosome 22. *Leukemia* **1988**, *2*, 760–762.

139. Grossman, A.; Mathew, A.; O'Connoll, M. P.; Tiso, P.; Distenfeld, A.; Benn, P. Multiple restriction enzyme digests are required to rule out polymorphism in the molecular diagnosis of chronic myeloid leukemia. *Leukemia* **1990**, *4*, 63–64.

140. Grossman, A.; Silver, R. T.; Szatrowski, T. P.; Gutfriend, A.; Verma, R. S.; Benn, P. A. Densitometric analysis of Southern blot autoradiographs and its application to monitoring patients with chronic myeloid leukemia. *Leukemia* **1991**, *5*, 540–547.

141. Negrin, R. S.; Blume, K. G. The use of the polymerase chain reaction for detection of minimal residual disease. *Blood* **1991**, *78*, 255–258.

142. Tkachuk, D. C.; Westbrook, C. A.; Andreeff, M.; et al. Detection of *bcr-abl* fusion in chronic myelogenous leukemia by in situ hybridization. *Science* **1990**, *250*, 559–562.

143. Erikson, J.; Griffin, C. A.; ar-Rushdi, A.; et al. Heterogeneity of chromosome 22 breakpoint in Philadelphia-positive (Ph[1]+) acute lymphocytic leukemia. *Proc. Natl. Acad. Sci. USA* **1986**, *83*, 1807–1811.

144. De Klein, A.; Hagemeijer, A.; Bartram, C. T.; et al. bcr rearrangement and translocation of the c-abl oncogene in Philadelphia positive acute lymphoblastic leukemia. *Blood* **1986**, *68*, 1369–1375.

145. Hermans, A.; Heisterkamp, N.; von Linden, M.; et al. Unique fusion of bcr and c-abl genes in Philadelphia chromosome positive acute lymphoblastic leukemia. *Cell* **1987**, *51*, 33–40.

146. Heisterkamp, N.; Knoppel, E.; Groffen, J. The first BCR gene intron contains breakpoints in Philadelphia chromosome positive leukemia. *Nucleic Acids Res.* **1988**, *16*, 10069–10081.

147. Denny, C. T.; Shah, N. P.; Ogden, S.; et al. Localization of preferential sites of rearrangement within the BCR gene in Philadelphia chromosome-positive acute lymphoblastic leukemia. *Proc. Natl. Acad. Sci. USA* **1989**, *86*, 4254–4258.

148. Chen, S. J.; Chen, Z.; Grausz, J. D.; et al. Molecular cloning of a 5' segment of the genomic phl gene defines a new breakpoint cluster region (bcr2) in Philadelphia-positive acute leukemias. *Leukemia* **1988**, *2*, 634–641.

149. Fainstein, E.; Marcelle, C.; Rosner, A.; et al. A new fused transcript in Philadelphia chromosome positive acute lymphocytic leukemia. *Nature* **1987**, *330*, 386–388.

150. Clark, S. S.; McLaughlin, J.; Crist, W. M.; Champlin, R.; Witte, O. N. Unique forms of the abl tyrosine kinase distinguish Ph[1]-positive CML from Ph[1]-positive ALL. *Science* **1987**, *235*, 85–88.

151. Kurzrock, R.; Shtalrid, M.; Romero, P.; et al. A novel c-abl protein product in Philadelphia-positive acute lymphoblastic leukaemia. *Nature* **1987**, *325*, 631–635.

152. Chan, L. C.; Karhi, K. K.; Rayter, S.; et al. A novel abl protein expressed in Philadelphia chromosome positive acute lymphoblastic leukemia. *Nature* **1987**, *325*, 635–637.

153. Walker, L. C.; Ganesan, T. S.; Dhut, S.; et al. Novel chimaeric protein expressed in Philadelphia positive acute lymphoblastic leukaemia. *Nature* **1987**, *329*, 851–853.

154. Secker-Walker, L. M.; Cooke, H. M. G.; Browett, P. J. Variable Philadelphia breakpoints and potential lineage restriction of bcr rearrangement in acute lymphoblastic leukemia. *Blood* **1988**, *72*, 784–791.

155. Collins, S. J.; Groudine, M. T. Chronic myelogenous leukemia: amplification of a rearranged c-abl oncogene in both chronic phase and blast crisis. *Blood* **1987**, *69*, 893–898.

156. Bartram, C. R.; De Klein, A.; Hagemeijer, A.; Carbonell, F.; Kleihauer, G.; Grosveld, G. Additional c-*abl/bcr* rearrangements in a CML patient exhibiting two Ph[1] chromosome during blast crisis. *Leukemia Res.* **1986**, *10*, 221–225.

157. Bartram, C. R.; Janssen, J. W,; Becher, R. Persistance of CML despite deletion of rearranged bcr/c-abl sequences. *Hematol. Bluttransfus* **1987**, *31*, 145–148.

158. Laneuville, P.; Sullivan, A. K. Clonal succession and deletion of *bcr/abl* sequences in chronic myelogenous leukemia with recurrent lymphoid blast crisis. *Leukemia* **1991**, *5*, 752–756.

159. Hagemeijer, A. E.; Smit, M. E.; Lowenberg, B.; Abels, J. Chronic myeloid leukemia with permanent disappearance of the Ph[1] chromosome and development of new clonal subpopulations. *Blood* **1979**, *53*, 1–14.

160. Reeve, A. E.; Morris, C. M.; Fitzgerald, P. H. Acquired homozygosity of the rearranged *bcr* allele during the acute leukemic phase of a patient with Ph-negative chronic myeloid leukemia. *Blood* **1988**, *72*, 24–28.

161. Ogawa, H.; Sugiyama, H.; Soma, T.; Massaocka, T.; Kishimoto, S. No correlation between locations of bcr breakpoints and clinical states in Ph[1]-positive CML patients. *Leukemia* **1989**, *3*, 492–496.

162. Przepiorka, D. Breakpoint zone of bcr in chronic myelogenous leukemia does not correlate with disease phase or prognosis. *Cancer Genet. Cytogenet.* **1988**, *36*, 117–122.

163. Jaubert, J.; Martiat, P.; Dowding, C.; Ifrah, N.; Goldman, J. M. The position of the M-bcr breakpoint does not predict the duration of the chronic phase or survival in chronic myeloid leukaemia. *Br. J. Hematol.* **1990**, *74*, 30–35.

164. Morris, S. W.; Daniel, L.; Ahmed, C. M. I.; Elian, A.; Labowitz, P. Relationship of the bcr breakpoint to chronic phase duration, survival and blast crisis lineage in chronic myelogenous leukemia patients presenting in early chronic phase. *Blood* **1990**, *75*, 2035–2041.

165. Tein, H. F.; Wang, C. H.; Chen, Y. C.; et al. Chromosome and bcr rearrangement in chronic myelogenous leukaemia and their correlation with clinical states and prognosis of the disease. *Br. J. Hematol.* **1990**, *75*, 469–475.

166. Tefferi, A.; Bren, G. D.; Wagner, K. V.; Schaid, D. J.; Ash, R. C.; Thibodeau, S. N. The location of chromosomal breakpoint site and prognosis in chronic granulocytic leukemia. *Leukemia* **1990**, *4*, 839–842.

167. Dreazen, O.; Berman, M.; Gale, R. P. Molecular abnormalities of bcr and c-abl in chronic myelogenous leukemia associated with a long chronic phase. *Blood* **1988**, *71*, 797–799.

168. Nowell, P. C.; Jackson, L.; Weiss, A.; Kurzrock, R. Historical communication: Philadelphia-positive chronic myelogenous leukemia followed for 27 years. *Cancer Genet. Cytogenet.* **1988**, *34*, 57–61.

169. Birnie, G. D.; Mills, K. I.; Benn, P. Does the site of the breakpoint in chromosome 22 influence the duration of the chronic phase in chronic myeloid leukemia? *Leukemia* **1989**, *3*, 545–547.

170. Mills, K. I.; Benn, P.; Birnie, G. D. Does the breakpoint within the major breakpoint cluster region (M-bcr) influence the duration of the chronic phase in chronic myeloid leukemia? An analytical comparison of current literature. *Blood* **1991**, *78*, 1155–1161.

171. Schaefer-Rego, K.; Dudek, H.; Popenoe, D.; et al. CML patients in blast crisis have breakpoints localized to a specific region of the BCR. *Blood* **1987**, *70*, 448–455.

172. Lee, M. S.; LeMaistre, A.; Kantarjian, H. M.; et al. Detection of two alternative *bcr/abl* mRNA junctions and minimal residual disease in Philadelphia chromosome positive chronic myelogenous leukemia by polymerase chain reaction. *Blood* **1989**, *73*, 2165–2170.

173. Morgan, G. T.; Hernandez, A.; Chan, L. C.; Hughes, T.; Martiat, P.; Wiedemann, L. M. The role of alternative splicing patterns of *BCR/ABL* transcripts in the generation of the blast crisis of chronic myeloid leukemia. *Br. J. Haematol.* **1990**, *76*, 33–38.

174. Lee, M.; Kantarjian, H.; Deisseroth, A.; Freireich, E.; Trujillo, J.; Stass, S. Clinical investigation of BCR/ABL splicing patterns by polymerase chain reaction (PCR) in Philadelphia chromosome (Ph[1]) positive chronic myelogenous leukemia (CML). *Blood* **1990**, *76*, 294a.

175. Dhingra, K.; Talpez, M.; Kantarjian, H.; et al. Appearance of acute leukemia-associated P190$^{BCR-ABL}$ in chronic myelogenous leukemia may correlate with disease progression. *Leukemia* **1991**, *5*, 191–195.

176. Lozzio, C. B.; Lozzio, B. B. Human chronic myelogenous leukemia cell line with positive Philadelphia chromosome. *Blood* **1975**, *45*, 321–334.

177. Collins, S. J.; Groudine, M. T. Rearrangement and amplification of c-abl sequences in the human chronic myelogenous leukemia cell line K-562. *Proc. Natl. Acad. Sci. USA* **1983**, *80*, 4813–4817.

178. Leibowitz, D.; Cubbon, R.; Bank, A. Increased expression of a novel c-*abl*-related RNA in K562 cells. *Blood* **1985**, *65*, 526–529.

179. Andrews, D. F.; Collins, S. J. Heterogeneity in expression of the *bcr-abl* fusion transcript in CML blast crisis. *Leukemia* **1987**, *1*, 718–724.

180. Daley, G. Q.; McLaughlin, J.; Witte, O. N.; Baltimore, D. The CML-specific P210 bcr/abl protein, unlike v-abl does not transform NIH/3T3 fibroblasts. *Science* **1987**, *237*, 532–535.

181. McLaughlin, J.; Chianese, E.; Witte, O. N. In vitro transformation of immature hematopoietic cells by the P210 bcr/abl oncogene product of the Philadelphia chromosome. *Proc. Natl. Acad. Sci. USA* **1987**, *84*, 6558–6562.

182. Young, J. C.; Witte, O. N. Selective transformation of primitive lymphoid cells by the BCR/ABL oncogene expressed in long-term lymphoid or myeloid cultures. *Mol. Cell. Biol.* **1988**, *8*, 4079–4087.

183. Daley, G. Q.; Baltimore, D. Transformation of an interleukin 3-dependent hematopoietic cell line by the chronic myelogenous leukemia-specific P210 bcr/abl protein. *Proc. Natl. Acad. Sci. USA* **1988**, *85*, 9312–9316.

184. McLaughlin, J.; Chianese, E.; Witte, O. N. Alternative forms of the *BCR-ABL* oncogene have quantitatively different potencies for stimulation of immature lymphoid cells. *Mol. Cell. Biol.* **1989**, *9*, 1866–1874.

185. Hariharan, I.K.; Harris, A. W.; Crawford, M.; et al. A *bcr-v-abl* oncogene induces lymphomas in transgenic mice. *Mol. Cell. Biol.* **1989**, *9*, 2798–2805.

186. Elefanty, A. G.; Hariharan, I. K.; Cory, S. *bcr-abl* the hallmark of chronic myeloid leukemia in man, induces multiple haematopoietic neoplasms in mice. *EMBO J.* **1990**, *9*, 1069–1078.

187. Daley, G.Q.; Van Etten, R. A.; Baltimore, D. Induction of chronic myelogenous leukemia in mice by the P210 *bcr/abl* gene of the Philadelphia chromosome. *Science* **1990**, *247*, 824–830.

188. Kelliher, M. A.; McLaughlin, J.; Witte, O. N.; Rosenberg, N. Induction of a chronic myelogenous leukemia-like syndrome in mice with v-*abl* and *BCR/ABL*. *Proc. Natl. Acad. Sci. USA* **1990**, *87*, 6649–6653.

189. Daley, G. Q.; Van Etten, R. A.; Baltimore, D. Blast crisis in a murine model of chronic myelogenous leukemia. *Proc. Natl. Acad. Sci. USA* **1991**, *88*, 11335–11338.

190. Szczylik, C.; Skorski, T.; Nicolaides, N. C.; et al. Selective inhibition of leukemia cell proliferation by BCR-ABL antisense oligodeoxynucleotides. *Science* **1991**, *253*, 562–565.

191. Vogelstein, B.; Fearon, E. R.; Hamilton, S. R.; et al. Genetic alterations during colorectal-tumor development. *N. Engl. J. Med.* **1988**, *319*, 527–532.

192. Vogelstein, A. Deadly inheritance (editorial). *Nature* **1990**, *348*, 681–682.

193. Sandberg, A.A. *The Chromosomes in Human Cancer and Leukemia*, 2nd ed. Elsevier, New York 1990, p. 479.

194. Fialkow, P. J.; Martin, P. J.; Najfield, V.; Penfold, G. K.; Jacobson, K. J.; Hansen, J.A. Evidence for a multistep pathogenesis of chronic myelogenous leukemia. *Blood* **1981**, *58*, 158–163.

195. Sagio, G.; Guerrasio, A.; Rosso, C.; et al. New type of *bcr/abl* junction in Philadelphia chromosome-positive chronic myelogenous leukemia. *Blood* **1990**, *76*, 1819–1824.

196. Soekarman, D.; van Denderen, J.; Hoefsloot, L.; et al. A novel variant of *bcr-abl* fusion product in Philadelphia chromosome positive acute lymphoblastic leukemia. *Leukemia* **1990**, *4*, 397–403.

197. Ishihara, T.; Sasaki, M.; Oshimura, M.; et al. A summary of cytogenetic studies in 534 cases of chronic myeloid leukemia in Japan. *Cancer Genet. Cytogenet.* **1983**, *9*, 81–93.

ADVENTURES IN *MYC*-OLOGY[†]

Paul G. Rothberg and Daniel P. Heruth

Advances in Genome Biology
Volume 3B, pages 337–414.
Copyright © 1995 by JAI Press Inc.
ISBN: 1-55938-835-8

†This chapter is a review of selected topics concerning the *myc* oncogene. It does not deal in any way with fungi.

INTRODUCTION

Why another review of the *myc* gene? This gene has been well reviewed in the past. General reviews have been published[1-4] and reviews of specific areas have been published in abundance.[5-16] On the other hand, the field moves so fast that there is a need for frequent updates. The literature on the *myc* gene is huge and fascinating. It is beyond the capabilities of the authors to cover the field in its entirety. In this review we will focus primarily on the involvement of this gene in abnormal growth. Thus, this paper will be a specific review of the literature on the involvement of the c-*myc* gene in disease. Related issues will be discussed, such as tissue specificity of the regulation of c-*myc* expression, and the clinical implications of alterations in the structure and expression of the c-*myc* gene. Whenever appropriate we will refer to other reviews for detailed background arguments and references to points which have been well covered in the recent literature. However, we will neglect some very interesting areas like the still very active research on the mechanisms of control of *myc* gene expression. We will also concentrate on the c-*myc* gene, neglecting the other members of the *myc* gene family. In Section II of this review, we will discuss the foundations of the study of the *myc* oncogene which were derived mostly from studies on fibroblasts and hematopoietic cells. In Section III, we will focus on other tissues which are less well understood, but extremely important in order to appreciate the importance of the c-*myc* gene in neoplasia.

II. CLASSIC *MYC*

A. The Basics

The human c-*myc* gene consists of three exons and covers about 5200 nucleotides. It is located on the q24 band of chromosome 8.[17-19] The entire gene and surrounding region have been sequenced.[20-22] There are two main promoters for transcription located 160 base pairs (bp) apart, and several other promoters which are used less frequently: one upstream of the gene and a cluster in the first intron.[21,23-25] Two main proteins are encoded by the c-*myc* gene: one starts at an AUG codon in exon 2, and the other uses the more unusual CUG codon for initiation in exon 1.[26] The protein products have estimated molecular masses between 62 and

68 kDa in human cells and are phosphorylated.[27–29] There is evidence for an additional protein product encoded by the first exon.[30–32] The main protein products of the c-*myc* gene are located in the nucleus of the cell and are loosely associated with chromatin, except during mitosis.[33–36] During mitosis, c-*myc* protein is more highly phosphorylated and it relocates throughout the cytoplasm.[34,37] Both c-*myc* RNA and protein are very unstable with estimated half-lives of 15–30 minutes.[28,29,38–41]

B. Ancient *myc*-Ology

Many of the now familiar concepts concerning the molecular biology of cancer were derived from the study of viral oncology. This includes a major contribution to the list of the dominantly acting transforming genes, called oncogenes. Although the study of the viral oncogenes today is not the dominant theme in the molecular biology of cancer, an understanding of the contributions from viral research is a necessary element in understanding any of the oncogenes. In this section we will briefly review the lessons from viral oncology with respect to the c-*myc* gene.

The *myc* gene was discovered in the avian acute transforming retroviruses CMII, MC29, MH2, and OK10.[42] The viruses in this group cause a broad spectrum of malignancies *in vivo*, including sarcomas, carcinomas, and myelocytomas, and also possess the ability to transform fibroblasts, epithelial cells, and bone marrow cells in culture.[43] As for all of the other retroviral oncogenes, the viral *myc* gene (v-*myc*) was derived from the host cell genome by a recombination event between a replication competent retrovirus and a preexisting host cell gene, in this case the cellular *myc* gene (c-*myc*). As a result of the recombination, the virus lost some of the functions required for its own propagation; thus, most of the acute transforming retroviruses replicate as two-virus systems with the transforming component containing the host cell derived oncogene, and a replication competent helper virus that supplies in *trans* the functions needed for viral propagation. As an example, the MC29 virus contains approximately 1600 bases of *myc*-specific sequences in a 5700-base genome.[44,45] The viral *pol* (reverse transcriptase) gene and parts of the *gag* and *env* genes were deleted. The *myc* gene is expressed as a fusion protein with the remaining portion of the *gag* gene.[46] The CMII, MH2, and OK10 viruses have different structures from MC29 and from each other; thus, they were born in separate *myc* transduction events.[47–52] A detailed description of the discovery, structure, and expressed oncoproteins of the v-*myc* family of avian retroviruses is contained in the review of Erisman and Astrin.[2]

How frequently do retroviruses arise with a *myc* oncogene in nature? As you will see in the remainder of this review, this event takes place frequently in laboratories. Infection with feline leukemia virus, a replication competent retrovirus that does not carry a host cell derived oncogene, has been found to result in the production of a recombinant provirus containing the cat c-*myc* gene in about 10% of the induced T-cell lymphomas.[53-57] Thus, the finding of several independent retroviruses with a *myc* oncogene has been seen in two species, chickens, and cats.

There is ample evidence that the v-*myc* sequences are responsible for the ability of MC29 to cause tumors. Three mutants of MC29 have been isolated that were capable of transforming fibroblasts *in vitro*, but were severely deficient in transforming macrophages, as compared to wild-type virus.[58] The mutants also had a severely lower pathogenic potential *in vivo*.[59] Analysis of the RNA from these mutants showed overlapping deletions between 200 and 600 bases in length.[60] The protein products of the mutant viruses were smaller than the wild-type 110-kDa *gag-myc* fusion protein, with molecular masses of 100, 95, and 90 kDa.[61] Studies of both RNA and protein demonstrated that the deletions occurred in *myc*-specific sequences and did not involve the *gag* domain.[60,61] Another group showed that a deletion in the 3′ portion of the v-*myc* gene caused a reduction in the ability of MC29 to transform fibroblasts.[62] In addition, the v-*myc* sequences in MH2 have been shown to be responsible for its ability to transform fibroblasts.[63] From these studies we can conclude that v-*myc* sequences are an important determinant in transformation and that separate functional domains of the v-*myc* gene may be involved in transformation of fibroblasts and macrophages.

We know that v-*myc* can cause cancer in chickens and that this gene was obtained from the host genome. This leads to two questions: (1) what is different about the v-*myc* gene that makes it a potent carcinogenic agent?, and (2) does the host cell c-*myc* gene cause cancer? We will answer the first question here and attempt to answer the second question in the remainder of this review.

A comparison of the nucleotide sequences of 4 avian v-*myc*s and chicken c-*myc* revealed several differences:[44,45,48,49,52,64-67]

1. The v-*myc* genes lack the introns and the noncoding first exon of c-*myc*.
2. Some of the v-*myc* genes were expressed as *gag–myc* fusion proteins.
3. Point mutations that change amino acids in the encoded protein have occurred in the v-*myc* genes.

These structural changes are also accompanied by a change in regulation, because the viral *myc* genes are under control of the viral regulatory elements and removed, in large part, from the host cell systems that normally regulate the expression of the c-*myc* gene. A number of experimental approaches have revealed that aberrant regulation of expression of the *myc* gene is a critical event in its activation to a carcinogenic agent. However, in comparing the biological potency of v-*myc* and c-*myc* there is some evidence for structural changes in v-*myc* that may be considered activating. Several biological assays can distinguish the transformation potency of chicken c-*myc* from the potency of the v-*myc* gene in MC29 and MH2 when expressed similarly.[68,69] A comparison of the activity of MC29 v-*myc* and chicken c-*myc* in their ability to transform myelomonocytic cells from mice revealed that point mutations in MC29 v-*myc* contributed to its transforming potential.[70] A mutation at position 61 of the chicken c-*myc* gene, a position that was found mutated in MC29, MH2, and OK10, activated its ability to transform fibroblasts.[71]

Is v-*myc* sufficient for transformation? We will return to this theme of biological potency, or spectrum of carcinogenic activity, of the *myc* gene repeatedly throughout this review as one of its central themes. The v-*myc* gene, as described earlier, is responsible for the ability of the MC29 virus to cause several malignancies *in vivo* and to transform cells in culture. However, the MC29 virus was a much less potent carcinogenic agent in Japanese quail than the MH2 virus.[72] This may be accounted for by the fact that MH2 has a second host cell derived oncogene, called *mil* or *mht*.[47,49] The mouse homolog of the *mil* gene (*raf*) was transduced by a murine retrovirus (murine sarcoma virus 3611) in which this gene is an oncogene in its own right. This story was further complicated by evidence that the *mil* gene does not contribute significantly to the biological activity of MH2.[63,72]

The ability of the v-*myc* gene to transform fibroblasts has been questioned. Although primary chick embryo and quail embryo fibroblasts that have been infected with MC29, or other members of the *myc* group of avian defective retroviruses, appeared transformed by several criteria, they were not tumorigenic in syngeneic animals or nude mice, respectively.[73,74] The v-*myc* gene derived from OK10, in a murine retroviral construct, was able to transform established and primary mouse fibroblasts by the criteria of anchorage independent growth, but the colonies were not as large as *ras* or *src* transformed cells, and tumorigenicity was not evaluated.[75] It seems that there are limitations to the biological potency of the *myc* oncogene.

C. Overexpression of c-*myc* is Carcinogenic: You Can Have Too Much of a Good Thing

Mice that lack both copies of the c-*myc* gene do not progress past day 10.5 of gestation.[76] Thus, *myc* null mutants are genetic lethals, and the *myc* gene is essential. Therefore, we are not going too far afield in calling the *myc* gene a good thing. However a tremendous body of experimental evidence has been amassed which supports the theory that overexpression of the *myc* gene contributes to malignancy. In this section, we will review some of the evidence for the importance of the control of expression of a nonmutated *myc* gene in hematopoietic cells and fibroblasts. In other sections we will review the evidence for other cell types.

The avian leukosis virus (ALV), which is a replication competent retrovirus that does not carry an oncogene, can cause bursal lymphoma in susceptible strains of chickens only after a long incubation. In most of the induced lymphomas an ALV regulatory element such as the promoter or an enhancer was found in the immediate vicinity of the c-*myc* gene and caused a big increase in its expression.[77,78] The chick syncytial virus, which is unrelated to ALV, caused bursal lymphomas with apparently similar integration events.[79] Similar events have been shown to occur in mouse T-cell lymphomas induced by replication competent murine retroviruses.[80–82] The viral integration events were shown to cause an increase in expression of the targeted gene. These viral systems serve to provide evidence that the host cell c-*myc* gene can become a carcinogenic agent when its normal regulation is disrupted, without having to be transduced by the virus. More detailed information on this type of oncogenesis can be found in several reviews.[2,83,84]

If a virus can integrate near the c-*myc* gene and activate its expression can a genetic rearrangement within a cell generate an activated *myc* gene without involving a virus? The answer to this question is yes, and was provided by the study of Burkitt's lymphoma and mouse plasmacytoma. In these diseases a chromosome translocation is commonly found which has been shown on a molecular level to result in the juxtaposition of the c-*myc* gene with an immunoglobulin gene.[18,85–88] The immunoglobulin genes are normally expressed in cells of the B-cell lineage, and the c-*myc* gene comes under the control of this active area of chromatin as the result of the translocation. The translocations themselves are probably the result of errors in the DNA rearrangements which generate functional immunoglobulin genes. The details of the rearrangements and the consequences for control of c-*myc* expression is an extensive story that has been well reviewed.[4,15]

To evaluate the effects of abnormally regulated c-*myc* expression on lymphoid cells, several groups have made transgenic animals with a c-*myc* gene linked to an immunoglobulin heavy-chain enhancer.[89–91] These transgenic mice and rabbits got lymphoid malignancies which were mono- or oligoclonal.[89–91] The fact that the tumors started from at most a few of the susceptible lymphocytes carrying the transgene indicated that other events must occur to get a malignancy besides deregulation of *myc* expression. In transgenic mice an early event was found to be a large expansion of actively growing pre-B lymphocytes.[92] The transgenic work confirms the implication from the findings in Burkitt lymphoma and mouse plasmacytoma that abnormal regulation of the c-*myc* gene is a carcinogenic event in lymphocytes.

Myeloid cells are also susceptible to transformation by c-*myc*. A recombinant retrovirus containing the c-*myc* gene was able to cause clonal monocyte–macrophage tumors in infected mice.[93,94] Mouse bone marrow cells infected *in vitro* gave rise to partially transformed cells that required CSF-1 for growth.[94]

The situation in fibroblasts is equally interesting. Increasing c-*myc* expression in established rodent fibroblasts by inserting an exogenous c-*myc* gene driven by a strong promoter caused the acquisition of tumorigenicity without causing gross morphological alterations.[95–97] In the experiments described in the last section which showed a lack of tumorigenicity in cells containing v-*myc*, primary cells were used as opposed to the established cell lines used here. The host cell is a critical variable in this type of experiment. Zerlin et al.[98] found that early and late passages of the same rat fibroblast cell line differed in the degree of transformation in response to forced expression of an exogenous c-*myc* gene. The cells which had been in culture for longer periods were more susceptible to transformation by c-*myc*.[98]

In primary rodent fibroblasts, the c-*myc* gene activated by way of strong expression, together with a Ha-*ras* gene activated by mutation, were necessary to get morphological transformation, while each gene alone was not sufficient.[99] Using this cotransformation assay, the level of expression of the c-*myc* gene was shown to be the most critical variable for demonstrating its potential carcinogenic activity.[100,101]

Another event which can result in enhanced expression of a gene is an increase in the number of genes, which is referred to as gene amplification. In the classic case, selection of cells for resistance to methotrexate often results in increased expression of dihydrofolate reductase, the enzyme which is inhibited by methotrexate, due to an amplification of the gene encoding this enzyme.[102] Similarly elevated expression of the c-*myc* gene can sometimes be attributed to an accompanying amplification of the gene. The first discovered example of c-*myc* gene amplification was in the human myeloid leukemia cell line HL60.[103,104] Throughout this chapter we will review the incidence of c-*myc* gene amplification in each malignancy covered. For most tumor types, amplification of the c-*myc* gene is an occasional or a rare finding. For the purposes of this section, however, *myc* gene amplification provides evidence for the importance of quantitative alterations in the expression of the *myc* gene.

Among tumors with a high level of expression of the c-*myc* gene some have amplification or rearrangement of the gene, but many have no detectable *cis* alteration (many references throughout); the elevated expression in these tumors is likely caused by a *trans* mechanism. The usual control mechanisms which regulate expression of the c-*myc* gene in the immortal, but nontumorigenic, mouse fibroblast cell line A31 were lost in its tumorigenic descendants without any detectable alteration in the encoding gene.[105] The expression of the c-*myc* gene in these cells was stuck at the high end of the normal transient increase seen in quiescent cells stimulated to proliferate.[105] This type of finding leads to use of the term *deregulated*—applied when an abnormally high level of expression of the c-*myc* gene is found in a tumor. The term, deregulated, is not quite correct since the gene is still under some type of control, but this control is not producing the same effect on expression of the gene as the regulatory mechanisms operative in the corresponding nonmalignant tissue. An interesting contribution to this area is the finding that the autoregulatory suppression of c-*myc* expression exerted by the c-Myc protein product is absent in many malignant cell lines.[106,107] It will be interesting when the *trans* deregulation of c-*myc* expression is understood to the extent of tracing the causality back to a carcinogenic mutational event.

As the (patient) reader will note in the second part of this chapter, the c-*myc* gene is expressed at an elevated level in some fraction of cases in many types of neoplasia. In some cases the reason for overexpression is obvious, but in many cases the reasons are yet unknown.

D. The *myc* Family

The c-*myc* gene, like most genes, has a family of related sequences. The N-*myc* gene was discovered because of its amplification in some neuroblastomas, and the L-*myc* gene was discovered because of its amplification in some small cell lung cancers.[108–110] Both genes have been shown to have biological activity similar to the c-*myc* gene.[111–113] Other members of the family are the B-*myc* and S-*myc* genes.[114,115] The B-*myc* gene does not have a DNA binding and dimerization motif as in c-, L-, and N-*myc*, but it does have a transcriptional activation domain.[116] The B-*myc* gene inhibited transformation by c-*myc* in the *ras* cotransformation assay.[116] The S-*myc* gene suppressed tumorigenicity in the rat neural tumor cell line RT4-AC.[114] However, this cell line has a *neu* gene activated by point mutation.[117] In view of the finding that overexpression of the c-*myc* gene suppressed the expression of the *neu* gene, and transformation induced by an activated *neu* gene, in NIH/3T3 cells,[118] the suppression of tumorigenicity by S-*myc* in the RT4-AC cells should not be interpreted as suggesting a generalized tumor suppressor function for the S-*myc* gene. The P-*myc* and R-*myc* genes are less well studied, but an initial report revealed transforming activity for the R-*myc* gene.[119]

E. What Does c-*myc* Do?

The c-*myc* gene, when expressed at a high level and removed from its normal control mechanisms, is carcinogenic. However, it must have some other function in control of cellular growth and differentiation besides causing cancer. In this section we will review the work which demonstrates a role for c-*myc* in cell growth and differentiation, a role in cell death, our present understanding of the biochemical function of its protein product, and its impact on response to cancer therapy.

Proliferation and Differentiation

What does *myc* do to a cell to make it transformed? Here we will look at the changes in proliferative behavior of cells in response to deregulated c-*myc* expression. Primary cells have a limited proliferative potential in culture. Elevated expression of a transfected c-*myc* gene has been associated with immortalization of primary fibroblasts.[100,120] In fact, the spontaneous immortalization of three different strains of rodent fibroblasts into immortal cell lines was accompanied by an increase in the expression of c-*myc* RNA.[121] However, the association of deregulated c-*myc* expression with tumorigenicity in established rodent fibroblast cell lines, which are already immortal, reveals that the immortalizing function of the c-*myc* gene is not the whole story.

The decision of a resting fibroblast to proliferate involves two steps: the acquisition of competence, which can be initiated by platelet derived growth factor (PDGF); and the decision to enter S phase and synthesize DNA, which can be induced by several agents called progression factors. The finding that the competence inducing growth factor PDGF also caused an increase in expression of the c-*myc* gene suggested that *myc* may mediate some or all of the effects of PDGF on cellular proliferation.[122] This has been tested in several ways:

1. Expression of a transfected c-*myc* gene or microinjection of Myc protein caused a reduction in the need for PDGF to get proliferation of rodent fibroblasts.[123,124]
2. Expression of an exogenous c-*myc* gene caused an increase in DNA synthesis without an increase in cellular proliferation in cells grown in a defined medium without additional progression factors.[125]
3. The response of several rodent fibroblast cell lines to stimulation by a progression type growth factor was enhanced by elevated expression of an exogenous *myc* gene.[126,127]
4. Forced elevated expression of the c-*myc* gene could mediate some of the aspects of competence, but did not completely replace PDGF.[123,127]

In a number of systems the transition of a resting cell to active proliferation is associated with an increase in c-*myc* expression that is usually transient. The association of PDGF with an increase in expression in fibroblasts was described above. Also in fibroblasts, the mitogens bombesin, transforming growth factor α, and epidermal growth factor caused an increase in expression of the c-*myc* gene.[128–131] In B-lineage lymphocytes c-*myc* expression was induced by the mitogens lipopolysaccharide, interleukin-7, anti-immunoglobulin antibodies, or an appropriate antigen.[122,132–134] In T lymphocytes, c-*myc* expression was induced by concanavalin A and monoclonal antibody OKT3 which reacts with the antigen receptor.[122,135] In erythropoietin sensitive cells of the erythroid lineage, erythropoietin caused an increase in expression of the gene.[136] In all cases examined, the response of the c-*myc* gene occurred before the cells actually began to make DNA. Other examples of this type of experiment will be described in later sections of this review. These experiments tell us that c-*myc* expression is associated with the decision to proliferate in many cell types.

The induction of differentiation *in vitro* is associated with a reduction in expression of the c-*myc* gene in many cell types including myeloid leukemia, erythroleukemia, myoblasts, teratocarcinoma, thyroid carcinoma, and colon carcinoma,[105,137–149] although there are a few exceptions, such as mouse keratinocytes, chicken lens epithelium, the human teratocarcinoma cell line Tera-2, and chronic lymphocytic leukemia cells.[150–153] The association of differentiation with lower c-*myc* expression in the majority of cases suggests that *myc* may have something to do with maintaining the dedifferentiated state. Direct evidence for this has been found in mouse preadipocyte, erythroleukemia, and myoblast cultures in which forced elevated expression of an exogenous c-*myc* gene blocked differentiation.[154–160]

Another line of experimentation which reveals the biological activity of the c-*myc* gene involves reducing the level of *myc* expression with antisense nucleic acids. The antisense nucleic acid has a sequence complementary to the c-*myc* mRNA and presumably binds to it and prevents translation, or possibly promotes degradation of the RNA by a double-strand specific ribonuclease. The antisense nucleic acid may be DNA or a variant which facilitates entry into the cell and resistance to nucleases. The addition of an antisense nucleic acid to a cell with a consequent reduction in expression of Myc protein has been seen to cause a reduction in proliferation of T lymphocytes, F9 teratocarcinoma cells, mouse erythroleukemia cells, human myeloid leukemia cell line HL60, human keratinocytes, a breast cancer cell line, COLO320 colon carcinoma cells, muscle cells, and *ras* transformed NIH/3T3 mouse fibroblasts.[161-171] In many cases, the growth arrest was accompanied by signs of differentiation.[162-165]

The bottom line of the experiments described or listed in this section is that the c-*myc* gene is an essential element in promoting cellular proliferation in many cell lineages. Control of *myc* expression is tantamount in these cells to control of growth. Deregulated expression of this oncogene also inhibits differentiation, either as a direct effect or as a consequence of promoting proliferation.

Apoptosis

The biological activity of the c-*myc* gene is usually connected with cellular proliferation. However, a body of evidence is accumulating which suggests a role for this gene in programmed cell death, a process called apoptosis. The sequence of events seen in this process involves a condensation of the cell nucleus and degradation of nuclear DNA to nucleosome sized fragments.[172] When either a mouse myeloid cell line or a rat fibroblast cell line were growth arrested and strong exogenous c-*myc* expression was induced, the cells underwent apoptosis.[173,174] The process was dependent on both growth arrest and expression of a functional c-*myc* gene.[173,174] In order for the *myc* gene to cause neoplastic transformation of a cell type that is susceptible to *myc* induced apoptosis, the process of apoptosis must be defeated. One mechanism for stopping apoptosis that was found to be effective in both fibroblasts and lymphocytes is the expression of the *bcl*-2 oncogene.[175,176] Another way to overcome apoptosis is to proliferate continuously, due either to a constant extracellular proliferative stimulus or another mutation in a growth controlling gene. The involvement of the c-*myc* gene in apoptosis may be a type of anti-cancer safety valve in which a cell that has lost control of its c-*myc* gene, but is not yet fully transformed, dies before another mutation(s) completes the process. On a more speculative note, this finding may provide one of the reasons why toxic agents, which cause a compensatory increase in cell division, are carcinogenic.[177] These agents may allow cells with deregulated *myc* expression to avoid apoptosis and expand in number. This would provide more time and more targets for further mutational events that complete the carcinogenic process.

Inhibition of c-*myc* expression using an antisense oligonucleotide blocked apoptosis in a T-cell hybridoma when the apoptosis was induced by activation of the CD3 T-cell receptor, but not when apoptosis was induced by dexamethasone.[178] Thus, c-*myc* expression is necessary for some pathways of apoptosis but not others.

The involvement of the c-*myc* gene in inducing apoptosis may explain a number of interesting experimental results. Expression of very high levels of Myc protein in Chinese hamster ovary fibroblasts appeared to be toxic.[179,180] The rat fibroblast cell line 208F, in which a highly expressed human c-*myc* gene had been introduced, produced rapidly growing tumors in immune-suppressed mice.[181] These tumors had a high mitotic rate, but also a high rate of apoptosis, compared with the parental cell line and tumors induced by expression of a Ha-*ras* gene.[181] Transformation of chicken embryonic bursal lymphocytes to a preneoplastic state with v-*myc* made the cells more susceptible to apoptosis induced by gamma radiation, or disruption of cell to cell contacts, than normal bursal lymphocytes.[182] Invasive lymphomas derived from v-*myc* transformed preneoplastic cells were resistant to the induction of apoptosis.[182]

Is the c-*myc* gene involved in apoptosis in epithelial cells? This question has not been answered yet; however, several experiments suggest that the answer may be yes. In transgenic mice in which c-*myc* expression was directed to the acinar cells of the pancreas an early event was increased apoptosis in the pancreas and smaller than normal organs in some mice.[183] In rat ventral prostate induced to atrophy by withdrawal of androgen, c-*myc* expression increased as the secretory epithelial cells underwent apoptosis.[184,185] Similarly, the expression of c-*myc* RNA increased in nude mouse xenografts of the estrogen-dependent breast cancer cell line MCF-7 during the apoptosis that preceded the regression of the tumor in response to withdrawal of estrogen.[186] In HeLa cells exposed to γ-interferon, expression of the c-*myc* gene was increased at 24–40 hours, which was prior to cell death by 72 hours.[187] A most recent example is the finding that there was an increase in c-*myc* RNA in nasopharyngeal carcinoma cells preceding apoptosis induced by vitamin K_3.[188]

Biochemical Aspects

The evidence for the protein product of the c-*myc* gene (Myc) being involved in DNA synthesis has been well reviewed.[1] The most recent work implicates Myc as a sequence specific DNA binding protein which controls the expression of other genes, possibly including itself. However, this evidence does not preclude the possibility that Myc is involved in DNA synthesis. The protein product of the c-*myc* gene is located in the nucleus.[33,35,36] The amino acid sequence of Myc contains the basic region helix-loop-helix and leucine zipper motifs which have been associated with sequence-specific DNA binding proteins.[8] The Myc protein has been shown to bind to the sequence 5'-CACGTG,[189–191] and is capable of activating transcription.[192–194] It binds as a heterodimer with a protein, called Max, which is required for transactivation of gene transcription *in vivo*.[194-196] Max can form homodimers with itself that bind to the same sites as the Max/Myc heterodimer.[197,198] However, the Max protein does not stimulate transcription. The Max/Max homodimer has the opposite effects as the Max/Myc heterodimer, perhaps by binding and not activating transcription in competition with the Max/Myc heterodimers which bind and enhance transcription.[193,194,198-202]

If Myc is a transcriptional activator, which genes are activated? (1) α-prothymosin, a nuclear protein associated with cell proliferation, (2) ornithine decarboxylase, (3) the p53 tumor suppressor gene, (4) the 70-kDa human heat-shock protein gene, (5) plasminogen activator inhibitor 1, (6) cyclin A, and (7) cyclin E were all activated by Myc.[203-208] For ornithine decarboxylase and p53, the activation by Myc required the CACGTG binding site, which in both cases was located in the first intron.[204,205] Thus, the effect of Myc on the p53 and ornithine decarboxylase genes is likely to be a direct effect.

Myc expression was also associated with decreased expression of some genes. Expression of the *neu*, cyclin D1, metallothionein I, HLA class I, LFA-1, two variants of histone H1, and several collagen genes have been shown to be decreased by Myc.[118,206,208-212] In the case of the collagen genes, a likely mechanism involves an alteration in the CCAAT transcription factor/nuclear factor 1.[213] It is not yet clear if all the genes whose expression is inhibited by Myc involve an indirect mechanism as is likely for the collagen genes.[213]

The Myc protein product has been implicated in control of the stability of intranuclear RNAs.[214,215] In particular, the plasminogen activator inhibitor 1 RNA in rat fibroblasts[207] and ribosomal RNAs in *Xenopus* oocytes[215] were stabilized. Does Myc induce the transcription of a protein(s) that controls the stability of these RNAs, or is there a direct effect of a Myc protein on RNA stability? Consistent with the latter is the intranuclear colocalization of Myc protein in the small nuclear ribonucleoprotein particles of rodent fibroblasts[36] and the nucleoli of *Xenopus* oocytes.[216]

Myc is unlikely to be the only factor regulating the genes listed above. Therefore, the finding that Myc is involved in regulation of a given gene in one experimental system may not be applicable in another cell type or state of differentiation in which the milieu of other factors regulating gene expression will be different.

Myc protein also regulates the expression of its own encoding gene.[217-221] Myc proteins have been shown to cause a decrease in transcription of the c-*myc* gene.[217,221] In some tumors with elevated c-*myc* expression the autoregulatory mechanism seems to be inoperative.[106,107] Some workers have found that Myc protein causes an increase in expression of the c-*myc* gene under certain conditions.[222,223]

Which of the putative Myc functions are responsible for the ability of the c-*myc* gene to transform cells? Regions in Myc that are essential for cellular transformation have been identified by directed mutagenesis and testing of the mutants using the *ras* cotransformation assay on primary rat fibroblasts, and/or transformation of established rat fibroblasts, while other regions have been identified as being dispensable for these biological activities.[224,225] The domains of Myc involved in heterodimer formation, sequence-specific DNA binding, transcriptional activation, and autoregulation were shown to be essential regions for transformation.[192,199,202,226] A correlation has also been demonstrated between the regions of the c-*myc* gene needed for inhibition of adipocyte differentiation and the regions needed for transformation of an immortalized rat fibroblast cell line.[227]

Cancer Therapy

Does the level of expression of the c-*myc* gene have an impact on the sensitivity of a cell to cancer therapy? In other sections of this chapter we will address the impact of alterations in the structure and expression of the c-*myc* gene on prognosis in various types of cancer. Here we will review the evidence, based on direct experiments, for a role of the c-*myc* gene in altering the way a cell responds to chemotherapy and radiation. Primary rat embryo cells, transfected with a mutationally activated Ha-*ras* gene, became slightly resistant to X-radiation compared to the parental cells, an effect which was enhanced by cotransfection of a v-*myc* gene, while the v-*myc* gene alone did not alter sensitivity to radiation.[228] Increased c-*myc* expression was also associated with resistance to γ-radiation in human fibroblasts derived from the skin of individuals with the Li-Fraumeni syndrome.[229] It is interesting that variant small cell lung cancer cell lines, which frequently have elevated c-*myc* expression, also are relatively resistant to X-rays.[230,231]

Another group inserted a c-*myc* construct, which was inducible by zinc ions due to a metallothionein promoter, into rat embryo cells and found that inducing increased c-*myc* expression increased the yield of cells having methotrexate resistance and amplification of the dihydrofolate reductase gene when the cells were cultured under selective conditions.[232] This result may be related to the finding that transfection with a construct that causes expression of a v-*myc* or c-*myc* gene in rat fibroblasts caused an increase in sister chromatid exchange and aneuploidy.[233] Unequal sister chromatid exchange is one of the events that can cause gene amplification.

Increased *myc* expression in NIH/3T3 cells, due to transfection of a construct with the c-*myc* gene under control of a retroviral promoter, was associated with an increase in resistance to several chemotherapeutic agents including cisplatin and adriamycin.[234] This result was not due to increased expression of the multiple drug resistance gene or the glutathione S-transferase-π gene.[234] Similar results were obtained using a murine erythroleukemia cell line in which increased expression of an exogenous c-*myc* gene caused increased resistance to cisplatin, while decreased c-*myc* expression caused by transfection of a c-*myc* antisense construct resulted in increased sensitivity to cisplatin.[235] The one discordant study in this group was the finding that two drug resistant variants of Chinese hamster lung cells had lost an amplification and overexpression of the c-*myc* gene which was present in the parental line.[236] Insertion of a construct that caused increased c-*myc* expression in one of these drug resistant variant lines resulted in an increase in sensitivity to several drugs such as vincristine and actinomycin D.[236]

III. GLOBAL *MYC*

The majority of human cancer strikes epithelial cells. In this section, we will look at several different tissues, mostly epithelial, to evaluate the frequency of occurrence of derangements in the structure and expression of the *myc* gene, the clinical correlates of such alterations, the biological consequences of intentionally deregulating c-*myc* gene expression, and the tissue-specific aspects of control of c-*myc* expression. Not every aspect will be covered with every tissue, mostly because not every aspect has been studied. One of the lessons learned by the authors is that despite the enormity of the literature on this gene there is still much that is unknown and in dispute.

A. Lung

Clinical Associations

The c-*myc* oncogene has been frequently found amplified in cell lines derived from small cell lung cancer (SCLC), particularly in a variant type that grows faster, has an increased ability to form colonies in semisolid medium, and lacks some of the differentiated neuroendocrine features found in classical SCLC, like expression of L-dopa decarboxylase.[230,237–243] In fact the association of c-*myc* amplification with the variant form of SCLC was one of the first such consistent associations of a *myc* gene alteration with a human carcinoma.[230] Amplification of the c-*myc* gene in a cell line established at relapse was a statistically significant negative prognostic indicator for the patient from whom the cell line was derived.[240]

In surgical specimens of SCLC the incidence of c-*myc* amplification has been found to be much lower; only 4 out of 122.[241–245] This discrepancy is probably due to the increased ease of establishing cell lines from tumors with amplification.[241] Amplification of the c-*myc* gene was found to be more common in tumors from patients who had been treated prior to biopsy.[241]

Overexpression of the c-*myc* gene in SCLC has been demonstrated on both the RNA and protein level.[230,238,243,246,247] Elevated expression of the gene has been demonstrated in the absence of gene amplification.[243,248] In one cell line this was shown to be due to increased initiation of transcription compared to SCLC lines that do not express the c-*myc* gene, and increased transcriptional elongation compared to a SCLC line that had both amplification and elevated expression of the gene.[248] In a series of 18 cell lines derived from small cell lung cancers there was a statistically significant correlation of increased expression of c-*myc* RNA with increased proliferative index, defined as the fraction of cells in the S, G_2, and M phases of the cell cycle.[249]

The c-*myc*-related genes, N-*myc* and L-*myc*, have also been found amplified in human SCLC.[110,250] In fact, the L-*myc* gene was discovered because of its amplification in SCLC cell lines and its sequence similarity to the c-*myc* gene.[110] The association of c-*myc* amplification with the variant form of SCLC was not seen with N-*myc* and L-*myc*.[110,250]

In nonsmall cell lung cancer the c-*myc* gene has been found to be overexpressed in some fraction and occasionally amplified or rearranged. Using only primary tumors, the c-*myc* gene has been found amplified in 4 out of 43 adenocarcinomas, 4 out of 28 squamous cell carcinomas, and in 2 out of 13 large cell carcinomas of the lung.[242,251,252] Immunohistochemical analysis of the expression of c-Myc protein in lung cancer showed an association between overexpression and poorly differentiated squamous cell carcinoma: 14 of 20 poorly differentiated tumors had elevated Myc protein compared with 6 out of 19 well and moderately differentiated tumors.[253] Using *in situ* hybridization, the association of strong c-*myc* expression with poorly differentiated squamous cell carcinoma was confirmed.[254] Other subtypes of lung cancer also had detectable c-Myc overexpression, including 13 of 22 adenocarcinomas.[253]

In giant cell carcinoma of the lung, two examples of c-*myc* rearrangement have been found: one in a cell line and one in a primary tumor.[255,256] In the primary tumor, a repeated DNA element of the L1 (Kpn) class was inserted upstream of exon 1.[257] In addition, one giant cell lung cancer cell line, C-Lu99, had an elevated level of c-*myc* RNA without apparent amplification or rearrangement of the encoding gene.[255]

Myc overexpression has been detected in nonmalignant tissue from lung cancer patients.[258] This may reflect a preneoplastic alteration in the lung tissue from which the tumor arose or, alternatively, growth factors produced by the tumor may cause increased *myc* expression in nearby nonmalignant tissue.

Bioassay

The significance of elevated expression and amplification of the c-*myc* gene in human lung cancer is supported by direct experiments. Transfection of the classic SCLC cell line H209 with a human c-*myc* gene, in order to increase its expression, resulted in clones with some similarities to the SCLC variants.[259] In particular, the transfected clones grew faster in culture and formed colonies in soft agar at an increased frequency compared to the parental cells, had a morphology like the variant, and a histology in nude mouse xenografts intermediate between the classic and variant lines.[259] Thus, at least some of the differences in phenotype between the classic and the variant SCLC cell lines can be attributed to differences in the expression of the c-*myc* gene.

Human bronchial epithelial cells immortalized by the SV40 large T antigen were made tumorigenic in nude mice by transfection of the c-*myc* and *raf* oncogenes.[260] Both oncogenes were required for tumorigenicity. The resultant tumors were described as "multidifferentiated," with markers of squamous, glandular, and neuroendocrine differentiation, similar to some small cell lung cancers seen *in vivo*.[261]

Regulation

In two experimental systems derived from epithelial cells of the lung the usual association of decreased c-*myc* expression with decreased proliferation was not observed. Type 2 epithelial cells (which line the alveoli) freshly isolated from neonatal rats grow in tissue culture in the presence of serum.[262] When serum was withdrawn, the cells stopped proliferating, but the expression of c-*myc* RNA remained constant. The same cell type isolated from adult rats did not proliferate in culture, but still had the same level of c-*myc* RNA as the proliferating neonatal cells.[262] This corresponds to the *in vivo* situation in humans in which the type 2 cells had the strongest expression of c-*myc* RNA in the alveolar region, without detectable expression of the Ki-67 antigen which is a marker of proliferating cells.[254]

In another system, the SCLC variant cell line NCI-N417 had a constant level of c-*myc* RNA when proliferating or when growth was inhibited by human recombinant leukocyte α-interferon, or the ornithine decarboxylase inhibitor difluoromethylornithine.[263] These agents have been shown to lower *myc* expression in other systems (see section on colon and rectum and Refs. 187, 264, and 265). Thus, epithelial cells of the lung may differ from other cell types in their regulation of c-*myc* expression. One possible explanation is that the c-*myc* gene is being regulated at a translational level,[262] although in a series of cell lines and xenografts derived from 23 small cell lung cancers, the expression of c-*myc* RNA was consistent with the level of c-Myc protein.[249] This area of investigation certainly deserves more attention.

B. Breast

Clinical Associations

Studies of the c-*myc* gene in human breast cancer have revealed amplification of the gene in several cell lines including SKBR-3, SW613-S, HBL100, VHB₁, and BSMZ.[266-271] This reflects the fact that the c-*myc* gene is amplified in many primary human breast carcinomas,[272-293] but there is a great deal of disparity in the frequency, ranging from 1% in a survey of 99 tumors[289] to 41% in a survey of 48 tumors.[281] A very large survey of 1052 human breast tumors showed 17.1% had amplification of the c-*myc* gene.[282] Studies of c-*myc* expression in human breast cancer have shown that the encoded RNA and protein is frequently expressed at a higher level in malignant than nonmalignant mammary tissue.[277,278,290,294-299] In tumors with amplification of the c-*myc* gene it was almost always expressed at an elevated level, but it was also overexpressed in breast tumors with no amplification or rearrangement.[277,278,290,294] The reason for elevated expression in breast cancers with a c-*myc* gene of apparently normal structure and quantity is unknown.

On a contrary note, one study found that cell lines made from normal human mammary epithelium expressed the c-*myc* gene at a higher level than a panel of breast cancer cell lines.[300] Whether this reflects a need for elevated *myc* expression in order to grow in culture or a normal consequence of growth stimulation of mammary epithelial cells is not known.

Rearrangement of the c-*myc* gene has been found in a small number of breast tumors.[279,281,283,301] The three rearrangements that have been characterized all involved the 3' portion of the gene,[279,283,301] but the significance of this unusual geography of *myc* rearrangement is unknown. In one interesting rearrangement a Line-1 repetitive element was found inserted into the second intron of the c-*myc* gene from an aggressive breast carcinoma, but was absent from nearby nonmalignant tissue.[283] This is not the only example of the mobilization of repetitive sequences that find the c-*myc* gene as a target. In canine transmissible venereal tumor a Line-1 element was found to be inserted 5' of the coding region.[302]

The search for prognostic indicators in breast cancer is particularly intense. A number of studies have attempted to link alterations in the structure and expression of many genes to clinical parameters. Amplification and/or overexpression of the c-*myc* gene has been correlated with increased age,[279] lack of progesterone receptors,[272,291] inflammatory carcinoma,[288] high S-phase fraction,[284] high proliferative activity as measured by expression of the nuclear antigen Ki-S1,[303] well differentiated tumors,[298] larger tumors,[285,299] and positive lymph nodes.[284,285,290,295,304] Most importantly, c-*myc* amplification has been found to be a negative prognostic indicator by a number of investigators.[272,273,277,284,285,304,305] Myc has also been found to be expressed at an elevated level in some specimens of fibrocystic disease of the breast, particularly in subtypes associated with proliferation and considered to be premalignant.[297,306] A correlation has been found between elevated expression of the c-*myc* gene in benign breast disease and a first-degree family history of breast cancer.[307]

Bioassay

The clinical studies provide evidence for an association between c-*myc* activation and neoplasia of the mammary epithelium. A number of studies have been done which provide direct evidence for the biological consequences of altering the control of expression of the c-*myc* gene in this tissue. In one study the human breast cancer cell line SW613-S, which has an amplification of the c-*myc* gene, was subcloned into several sublines that differed with respect to the degree of *myc* amplification and expression.[308,309] The sublines with 30- to 60-fold amplification were tumorigenic in nude mice, while the sublines with 2- to 3-fold amplification were not tumorigenic. When the nontumorigenic cells were transfected with a c-*myc* gene under the influence of the powerful SV40 early region enhancer, they gained the property of tumorigenicity.[308,309] In a similar experiment a c-*myc* coding region under the control of a retroviral enhancer/promoter altered the properties of an immortalized nontumorigenic mouse mammary epithelial cell line so that it acquired the ability to grow in soft agar and produce tumors in nude mice.[310] Another nontumorigenic mouse mammary epithelial cell line did not gain the ability to produce tumors when it was transfected with a v-*myc* gene under control of a similar retroviral enhancer/promoter, although it did gain other properties such as a limited ability to grow in soft agar and altered responsiveness to lactogenic hormones.[311] As seen below, an activated c-*myc* gene alone cannot produce a mammary carcinoma. Thus, the apparent discrepancy between the two latter transfection experiments probably lies in the genomes of the host cells used. This type of discrepancy is reminiscent of the findings from studies in fibroblasts in which the apparent biological activity of the c-*myc* gene was strongly dependent on the host cell.

Among the most convincing work comes from transgenic mouse experiments. Two groups have produced mice with c-*myc* transgenes that are expressed in mammary epithelium: In one set of experiments the c-*myc* gene was linked to the transcriptional control machinery of mouse mammary tumor virus (MMTV) which is inducible by glucocorticoids,[312] and in the other the c-*myc* gene was linked to the promoter and upstream region of the whey acidic protein (WAP) gene which is normally expressed in lactating mammary gland.[313] Both transgenics produced an excess of adenocarcinoma of the mammary epithelium, but only after one or more pregnancies.[312-314] In the transgenic mice that expressed the WAP-*myc* construct, 80% got palpable tumors after two lactation periods.[313] The tumors expressed the transgene as well as the endogenous WAP gene and the β-casein gene, which is another lactation specific gene. This indicates that an event(s) occurred which caused deregulation of the expression of the lactation-specific genes, together with the c-*myc* transgene, all of which were under control of similar regulatory elements. In the MMTV-*myc* transgenic mice, only about 10% of the mice developed mammary cancer, again only after one or more pregnancies, indicating a need for hormonal stimulation of the transgene in order to get neoplasia. In both transgenic systems the tumors grew out from nonmalignant mammary epithelium which expressed the transgene; thus, abnormal c-*myc* expression by itself was not sufficient to cause neoplasia.

When the *myc* transgenic mice were crossed with mice containing a Ha-*ras* gene activated by point mutation (MMTV/v-Ha-*ras* or WAP/human mutant Ha-*ras*), double transgenics were generated that developed adenocarcinoma of the mammary epithelium with a very high frequency. Close to 90% of the MMTV double transgenics developed breast cancer, including the males.[315] In the WAP double transgenics all the females developed multiple mammary tumors after several pregnancies, and normal mammary epithelium failed to develop, so these double transgenics were unable to nurse.[316] In both cases nonmalignant mammary epithelium that expressed both the *ras* and *myc* transgenes were found, indicating that additional steps were still required to get a fully malignant adenocarcinoma of the breast.[315,316]

In the transgenic mouse experiments all of the cells in the mammary epithelium carried the transgene. In nature, single cells that acquire mutations are surrounded by normal cells which may have a suppressive effect on growth of the premalignant cell.[69,317,318] In order to model this situation in breast carcinogenesis, mammary epithelial cells were manipulated in short-term primary culture, then implanted into the cleared mammary fat pads of immature mice.[319] In this system normal cells grow into a normal gland. When the cells were exposed to a recombinant defective retrovirus containing a v-*myc* gene, only some of the cells got infected and contained the activated *myc* gene. Transplants of these cells produced mammary glands with areas of hyperplasia, but no neoplasia, unlike the transgenic mouse models in which an activated c-*myc* gene can produce an increase in mammary neoplasia.[319] Similar experiments with v-Ha-*ras* produced no tumors, but infection with both v-Ha-*ras* and v-*myc* produced mammary tumors in the majority of animals.[320] Thus, the transplantation experiments support the transgenic mouse and DNA transfection experiments in providing a role for the *myc* gene in mammary carcinogenesis.

Infection of newborn mice with a recombinant murine retrovirus containing the MC29 v-*myc* gene resulted in a variety of tumors including adenocarcinoma of the mammary gland.[321] These tumors were clonal and did not develop immediately after infection, which indicates that other events are probably needed to get a malignancy out of a cell which expresses a high level of v-*myc*.[321] These experiments support the conclusion that an activated *myc* gene is a carcinogenic agent in mammary epithelial cells, but it is not sufficient to produce a tumor alone.

Regulation

The level of c-*myc* RNA in cultured mammary carcinoma cell lines which are dependent on estrogen for growth, like MCF-7 and T47D, was low when the cells were in a quiescent state induced by deprivation of estrogen, but was transiently stimulated by estrogen.[268,322–324] The effect was shown to be due to increased transcription of the c-*myc* gene.[323] Exposure to an antisense oligonucleotide that inhibits expression of the c-*myc* gene, inhibited the proliferative response of MCF-7 cells to estrogen.[168] This indicates that the estrogen stimulation of c-*myc* gene expression is a necessary step in estrogen-dependent proliferation. Breast cell lines that are not estrogen-dependent (e.g., MDA-MB-231, BT-20, and HBL-100) have a constitutively high level of c-*myc* expression.[268,323] This may mean that deregulation of c-*myc* expression is a step in going from an estrogen-dependent to an estrogen-independent tumor. However, when an activated c-*myc* gene was trans-fected into MCF-7 cells, estrogen was still required for tumorigenicity in nude mice.[325] Thus, deregulation of *myc* expression was not sufficient to create a fully estrogen-independent phenotype. The effects of estrogen on growth of mammary epithelial cells involves more than just stimulation of expression of the c-*myc* gene.

Tamoxifen is an agent which blocks the growth effects of estrogen. In MCF-7 and T47D cells tamoxifen caused a decrease in the level of c-*myc* RNA.[268,322,326] The anti-estrogen ICI 164,384, as well as tamoxifen, blocked the effect of estradiol on increasing the expression of c-*myc* RNA in MCF-7 cells.[327] A similar result was obtained from an *in vivo* experiment in which breast cancer biopsies from women pretreated with tamoxifen had significantly less c-*myc* RNA, measured by *in situ* hybridization, than a similar population with breast cancer that was untreated before biopsy.[328]

In vivo, progesterone is believed to promote the growth of mammary epithelium, even though it inhibits the growth of most breast cancer cell lines including T47D. The synthetic progestins medroxyprogesterone acetate (MPA) and ORG 2058 caused a rapid transient increase in expression of c-*myc* RNA in T47D cells.[326,329] However, the initial effect of MPA was to cause a slight increase in the number of cycling cells, so this does not contradict the usual finding of an increase in c-*myc* expression associated with the transit of cells from a quiescent to a cycling state. Interestingly, there is an apparent contradiction to this association when the agents interferon-γ or epidermal growth factor caused an increase in the level of *myc* RNA in the MDA-468 breast carcinoma cell line even though they decreased the growth of these cells.[330,331]

C. Cervix, Ovary, and Uterus

Clinical Associations, Cervix

The c-*myc* gene has been found to be frequently amplified in carcinoma of the uterine cervix.[10,332] Amplification was found more frequently in advanced stage III and IV tumors (49%) than in early stage I and II tumors (6%).[10,333] In a survey done in China, no amplification of the c-*myc* gene was seen in the 17 patients studied.[334] It is not clear if this was due to geographic differences in the pathology of the disease or to the collection of earlier stage tumors. Although almost all tumors with amplification of the c-*myc* gene had elevated expression of its RNA, only about 25% of tumors with elevated expression had amplification;[333] thus the reason for elevated expression of the c-*myc* gene in these tumors is not yet explained. The gene was similarly overexpressed in a number of cervical carcinoma cell lines, also without amplification or rearrangement.[334]

An examination of archival biopsies of cervical cancers using flow cytometry of nuclei showed a higher level of Myc protein in nonmalignant cervix than in most tumors.[335] The apparent contradiction to the studies showing high levels of c-*myc*

RNA and frequent gene amplification is unresolved. However, an immunohisto-chemical approach using the same monoclonal antibody revealed elevated but variable expression of Myc protein in 11 invasive tumors compared to no staining in six specimens of normal cervical epithelium.[336]

In advanced tumors there was no prognostic significance to Myc protein expression detected by immunohistochemistry.[337] However, in early invasive cervical carcinoma, an elevated level of c-*myc* RNA was shown to be a strongly negative prognostic indicator.[333,338,339] In a study of 93 patients, multivariate analysis revealed that greater than 5-fold elevated expression of the c-*myc* gene gave a 3.7-fold greater risk of relapse.[333,339]

Clinical Associations, Uterus

Amplification of the c-*myc* gene has been found in endometrial cancer: 10 out of 16 tumors in one survey,[340] and 1 out of 3 in another.[341] Amplification was associated with advanced stage, lack of differentiation, and serous papillary adenocarcinoma histology.[340,341]

Clinical Associations, Ovary

The c-*myc* gene was also found amplified in about one-quarter of the ovarian carcinomas that have been reported.[342–346] Amplification had no prognostic value in one survey of 17 tumors,[342] while in another survey of 16 adenocarcinomas c-*myc* amplification was associated with poor differentiation, high number of mitoses, and a high degree of nuclear atypia.[343] *myc* RNA was shown to be overexpressed in more than one-third of human ovarian tumors,[347–349] with stage III serous adenocarcinomas having an 84% incidence of overexpression.[348] In a study of radiation-induced ovarian carcinomas in mice, 25% (7/28) had overexpression of c-*myc*, but no amplification or rearrangement was seen.[350] Overexpression of c-*myc* RNA was associated with progression of disease in a survey of 28 stage III and IV human ovarian carcinomas.[351] Myc protein was expressed in the tumor cells, not stromal elements.[348] A study of c-Myc protein by flow cytometry of nuclei from archival specimens of serous papillary ovarian cancer revealed that about two-thirds had a c-Myc protein signal twofold or more above the median signal from normal ovary.[352] In mucinous tumors of the ovary, malignancy was associated with the presence of c-Myc protein in the cytoplasm and the nucleus, instead of the more usual finding of an exclusively nuclear location.[353]

Regulation

The mammalian uterus and the avian oviduct are estrogen-responsive tissues, which respond with cellular proliferation. Estrogen caused an increase in expression of c-*myc* RNA in the prepubertal rat uterus[5,354–356] and the avian oviduct.[357]

Table 1. Differentiation Agents which Cause a Reduction in c-*myc* Expression in Ovarian and Cervical Cancer Cell Lines

Cell Line	Agent	Reference
HOC-7, ovarian cancer	Dimethylsulfoxide	361
	Dimethylformamide	
	Retinoic acid	
	Transforming growth factor-β1	
HeLa, cervical carcinoma	Interferon-γ	363
	Tumor necrosis factor	
HeLa, cervical carcinoma	Dimethylsulfoxide	362

Progesterone, which inhibits estrogen-induced proliferation in the uterus, caused a transient reduction in the level of c-*myc* RNA in estrogen-stimulated chicken oviduct.[358] Vanadate caused a transient increase in expression of c-*myc* RNA in CaOv ovarian carcinoma cells.[359] Vanadate is a tyrosine phosphatase inhibitor that is mitogenic in many cells. This experiment implicates tyrosine phosphorylation in the control of c-*myc* expression in an epithelial cell.

Dexamethasone, a glucocorticoid, causes inhibition of growth of the avian oviduct and also caused a transient drop in the expression of c-*myc* RNA.[360] The ovarian cancer cell line HOC-7, and the cervical carcinoma cell line HeLa, responded to several agents (see Table 1) with a concomitant reduction in both growth and level of c-*myc* gene expression.[361–363] Thus, the tissues of the female reproductive tract are consistent with the general trend of c-*myc* expression increasing in response to growth stimulation and decreasing in response to growth inhibition.

Tumor necrosis factor (TNF) and interferon-γ (IFN-γ) both reduced the expression of the c-*myc* gene in HeLa cells by reducing transcription.[363] Interestingly, the effects of these two agents followed different pathways: The effect of IFN-γ was inhibited by the protein synthesis inhibitor cycloheximide, which indicates the need for the synthesis of a new protein to lower the expression of c-*myc* RNA, while the effect of TNF was not inhibited by cycloheximide, indicating a more direct pathway. Addition of saturating amounts of each agent together resulted in an additive effect; thus, the rate limiting step for each agent is different.[363]

D. Brain

Clinical Associations

Medulloblastoma is a rare type of brain cancer that is predominately pediatric. The c-*myc* gene was found to be expressed at an elevated level in the majority (4/7) of medulloblastomas that have been reported.[364,365] Amplification of the c-*myc* gene has been found frequently in cell lines and xenografts derived from medulloblastomas, but much less frequently (less than 10%) in primary tumors.[364,366–371] The difference between primary tumors and cell lines in amplification of the c-*myc* gene is likely accounted for by the outgrowth in culture of an undetectably small fraction of cells which did have amplification in the original tumor.[367] Medulloblastomas are capable of metastasis, and it will be interesting to see if some metastatic medulloblastomas are different from the original tumor with respect to *myc* amplification. Rearrangement of the c-*myc* gene was reported in one medulloblastoma,[367] and mutation of the first exon was reported in another.[365]

Amplification, rearrangement, and elevated expression of the c-*myc* gene was found in a cell line derived from a glioblastoma multiforme.[372] However, the c-*myc* gene was not found amplified in patient specimens of 81 other glial tumors.[364,366,373] Expression of the c-Myc protein was shown to be elevated in glial tumors and glial tumor cell lines over that in normal brain and nonmalignant glial cell lines, with an increased incidence of overexpression in higher grade tumors.[374,375]

A malignant meningioma cell line has been established with amplification of the c-*myc* gene.[376] In a study of 19 primary tumors, the expression of c-*myc* RNA was 5-fold or more above nonmalignant brain tissue in 12, but none of the tumors in this series had amplification or rearrangement of the encoding gene.[377]

There was a 10-fold decrease in the level of c-*myc* RNA between newborn and 3-week-old whole mouse brain.[378] This decrease was shown to be due to a complex set of regulatory events involving decreased transcriptional elongation and initiation as well as posttranscriptional events.[378] The contribution of different tissues to the results in whole brain may contribute to the apparent complexity. *In situ* analysis of c-*myc* expression during development of the human and mouse brain revealed increased expression of c-*myc* RNA in both actively proliferating regions, and differentiating postmitotic zones.[379,380] For example, on the 10th day after birth of the C57BL/6 mouse, c-*myc* was found in the actively proliferating external granular layer of the cerebellum, as well as in the Purkinje cells which were differentiating and no longer proliferating at this stage.[380]

Bioassay

Several experimental systems have been established which demonstrate the biological activity of a strongly expressed exogenous c- or v-*myc* gene in cells of the central nervous system. MC29, the prototypical avian retrovirus with a *myc* oncogene, has been shown to induce transformation in neuroretina cells from 7-day-old chicken embryos and in neural crest cells derived from 2-day-old quail embryos.[381,382] In both cases differentiation was not inhibited: The chick neuroretina cells produced both neural and glial cell types,[381] while the neural crest cells expressed catecholaminergic traits.[382] The latter result was also obtained after infection with a recombinant retrovirus expressing a c-*myc* oncogene.[382] Recombinant retroviruses with the v-*myc* and v-*raf* oncogenes, or a c-*myc* oncogene alone, have been shown to immortalize primary cultures of murine microglial cells and neuroepithelial cells from the mesencephalon of mouse embryos, respectively.[383,384] The immortalized microglial cells retained a number of differentiated functions including phagocytosis,[383] and one of the immortalized neuroepithelial cell lines was capable of differentiating to glial or neural cells.[384]

In a transgenic mouse experiment, two transgenic lines were crossed, one with a human T-cell leukemia virus (HTLV) *tax* gene driven by an immunoglobulin promoter, and the other line with a c-*myc* gene driven by a HTLV promoter, which is *trans*-activated by the *tax* gene product.[385] Mice with both transgenes were produced which had a 76% incidence of brain tumors by 90 days. The histology of the tumors most resembled the human primitive neuroectodermal brain tumors (PNET), a tumor type which includes the medulloblastoma. It is possible in this transgenic experiment that the *tax* gene had a role both in activating the *myc* transgene and also in activating some other cellular gene(s) in order to get such a high rate of tumorigenesis.[385]

In a cell transplant type of approach, fetal rat brains were removed, made into a single cell suspension, and infected with recombinant retroviruses, then reinjected into the cranium of an adult syngeneic rat.[386] One of 13 transplants that had been infected with a v-*myc* retrovirus developed a PNET from the transplanted cells. This is the same type of tumor described earlier in the transgenic experiment.[385] When the transplanted cells had been infected with a recombinant retrovirus that expressed both v-*myc* and v-Ha-*ras*, all the rats developed, within 2–4 weeks, multiple highly malignant tumors some of which eventually expressed glial fibrillary acidic protein, a marker of glial cells. Infection with *ras* alone gave tumors in only half the recipient rats with a much longer latency.[386] Thus, several lines of experimentation reveal a potent biological activity for the *myc* gene in tumorigenesis of both neural and glial cells in the central nervous system.

E. Prostate

Clinical Associations

Expression of c-*myc* RNA was elevated 5-fold over normal prostate in 4 out of the 7 human prostatic carcinomas that were analyzed in one study,[387] and 2-fold or more over the level of *myc* RNA in benign prostatic hypertrophy in 10 out of 12 prostatic carcinomas in another.[388] *myc* RNA levels were also elevated in benign prostatic hypertrophy, but not as much as in the carcinomas.[387] Using *in situ* hybridization, c-*myc* RNA was localized to the prostatic epithelial cells in benign prostatic hypertrophy.[389] Growth in nude mice of the PC-3 prostatic carcinoma cell line resulted in tumors with a 10- to 12-fold amplification of the c-*myc* gene; however, amplification of the c-*myc* gene was not detected in the parental cell line.[390] Presumably an undetectably small fraction of cells with amplification of the c-*myc* gene in the original cell line had a selective advantage for growth in nude mice.[390] There was no correlation between expression of c-Myc protein, detected by immunohistochemistry, and prognosis, in a survey of 45 early (stage A1) adenocarcinomas of the prostate.[391]

Bioassay

The biological activity of the c-*myc* oncogene in the epithelial cells of the prostate has been demonstrated in an organ reconstitution experiment in which fetal cells from the urogenital sinus of a C57BL/6 mouse were infected with retroviral vectors containing v-Ha-*ras*, v-*myc*, or both, and then reimplanted into the renal capsule of syngeneic hosts.[392] The *ras* oncogene induced dysplasia; the *myc* oncogene induced hyperplasia; and the cells infected with the combination of both *ras* and *myc* developed adenocarcinomas. The tumors were clonal, and not all cells that expressed the oncogenes formed tumors. Thus, although the combination of *ras* and *myc* were necessary for tumorigenesis, they were not sufficient.[392] When this experiment was done using cells from BALB/c mice, instead of C57BL/6, the result of infection with *ras* + *myc* was focal epithelial hyperplasias and very rarely carcinomas.[393] This result emphasizes the importance of the host cell when evaluating biological activity. Four cell lines made from the hyperplasias induced by v-Ha-*ras* in the C57BL/6 mice all had mutations in the p53 gene, while nine cell lines made from the adenocarcinomas induced by v-Ha-*ras* + v-*myc* all lacked mutations in the p53 gene.[394] The activated *myc* gene seems to have made p53 mutations unnecessary in this system.

Regulation

The prostate depends on androgen for growth and maintenance of the adult organ. Castration results in atrophy of the ventral prostate accompanied by the death of numerous epithelial cells. Castration of the rat resulted in a paradoxical increase in expression of c-*myc* RNA, which was due to the regressing epithelial cells.[184,185] Administration of androgen reversed the increase in c-*myc* RNA.[184] This system seems to be an *in vivo* manifestation of the c-*myc* function involved in apoptosis. This increase in *myc* expression is probably directly involved in the programmed death of the prostatic epithelial cells when deprived of androgen.

After castration, readministration of androgen causes a regrowth of the regressed prostate. In rats that had been castrated 7 days previously androgen caused a transient increase in the expression of c-*myc* RNA.[395] Insight into the regulation of c-*myc* gene expression by androgen was gained from a model system in which mibolerone, a synthetic androgen which inhibits growth of the prostate carcinoma cell line LNCaP, caused a decrease in expression of c-*myc* RNA due to a decrease in its synthesis.[396] Thus, *myc* appears to be involved in both proliferation and apoptosis of prostatic epithelial cells.

It is hard to reconcile the latter experiments showing the usual association of increased c-*myc* expression and proliferation, and the earlier described experiments showing a role for expression of the *myc* gene in prostate carcinogenesis, with the experiments showing increased c-*myc* expression in cells undergoing apoptosis. Other factors are involved in the decision to proliferate or die in cells that have increased expression of the c-*myc* gene. This point was illustrated by experiments using two transplantable adenocarcinomas of the prostate that were produced in the organ reconstitution experiments described above.[392] An increase in *myc* expression in response to androgen deprivation was seen in both carcinomas, one of which was androgen sensitive and another that was androgen insensitive for growth in syngeneic mice.[397,398] However, the androgen sensitive cells also induced a gene associated with apoptosis of the prostate, while the androgen insensitive tumor cells did not induce this same gene in the castrated host.[398] It seems that expression of the c-*myc* gene is necessary, but not sufficient, for both apoptosis and proliferation in the prostate.

F.   Testes

Clinical Association

The protein product of the c-*myc* gene was expressed at an elevated level in testicular cancer compared to normal testes.[399] Seminomas and teratomas with intermediate differentiation and yolk sac elements had the highest levels of c-Myc protein.[399,400] Interestingly, increased expression of the c-*myc* gene was correlated with a better prognosis in testicular cancer![400]

Bioassay

A transgenic mouse line, MMTV-*myc*, which expressed the c-*myc* transgene in a variety of tissues, including testes, developed Sertoli cell testicular neoplasms in four of 35 mice studied.[314]

Regulation

In mouse testes the association between rapid growth and elevated expression of the c-*myc* gene did not hold for spermatogonia which actively proliferate and have very low levels of c-*myc* RNA.[401] Other cell types in the testes, such as Leydig cells and Sertoli cells, expressed abundant c-*myc* RNA when actively proliferating.[401]

Leydig cells from the mouse and the pig *in vitro*, and the rat *in vivo*, responded to human chorionic gonadotropin, which promotes proliferation of this cell type with a transient increase in expression of the c-*myc* gene.[402–404] The freshly isolated pig Leydig cells also responded to the progression factors epidermal growth factor and basic fibroblast growth factor with an increase in the level of c-*myc* RNA.[403] The rate-limiting step for the stimulation of c-*myc* gene expression by chorionic gonadotropin differs from that for epidermal growth factor and basic fibroblast growth factor.[403]

G. Kidney

Clinical Associations

The c-*myc* oncogene has been found overexpressed with a high frequency in human renal cell carcinomas,[405–407] and the cell lines derived from this malignancy.[408] An association was found between elevated c-Myc protein, detected by immunohistochemistry and nuclear pleomorphism.[405] It was also expressed at an elevated level in kidney tumors induced by estrogen in Syrian hamsters.[409] The c-*myc* gene was not found amplified or rearranged in any of the six human renal cell carcinomas with an elevated level of c-*myc* RNA that were analyzed.[406] A family with inherited renal cell carcinoma that cosegregated with a chromosomal translocation [t(3;8)(p14.2;q24.1)], which relocated the c-*myc* gene to chromosome 3, was found to have its chromosome 8 breakpoint at least one-half million bp distant from the gene.[410] Thus, it is doubtful that the translocation directly caused an alteration in the regulation of the c-*myc* gene; however, very long-range effects cannot be completely dismissed.

The analysis of the structure and expression of the *myc* gene in renal malignancies is still very incomplete. However, polycystic kidney disease holds a fascinating story on the potential impact of deregulation of *myc* expression in a nonmalignant proliferative disease. The polycystic kidney diseases are a collection of disorders which are characterized by the growth of a multitude of cysts, lined with epithelial cells, that cause enlargement and eventual failure of the kidney. In humans the most common type is an adult onset autosomal dominant inherited condition. However, there is a rare autosomal recessive form of the disease with a very early onset and a mouse model. In autosomal recessive polycystic kidney disease of the C57BL/6J mouse, the c-*myc* oncogene has been shown to be highly overexpressed; 3-week-old affected whole kidneys had 25- to 30-fold overexpression compared with normal kidney of the same age.[411,412] The c-*myc* gene was shown to be expressed in proliferating cells in the normal developing mouse kidney, but by 3 weeks of age its expression was greatly decreased.[413,414] However, in the polycystic kidneys c-*myc* RNA was increasingly expressed in the cells lining the cysts, and in the 3-week-old animals even the apparently uninvolved proximal tubules expressed more c-*myc* RNA than in normal kidney.[414] The expression of c-*myc* RNA increases with proliferation in adult mouse kidney (see below), but 3-week-old polycystic mouse kidneys were not in a state of very active proliferation.[411,412] It seems that the c-*myc* gene may play a role in the pathogenesis of this disease. Perhaps the gene for autosomal recessive polycystic kidney disease is a negative regulator of c-*myc* gene expression, that is specific for kidney epithelial cells.

Bioassay

One important question is: Can deregulated c-*myc* expression, by itself, cause polycystic kidney disease? This question was answered by accident in a transgenic mouse experiment in which the c-*myc* gene was linked to the promoter from the β-chain of hemoglobin, and an SV40 enhancer, with the intention of overexpressing the *myc* gene in cells of the erythroid lineage.[415] The gene was instead expressed in epithelial cells of the kidney, probably because of the creation of unexpected signals when the globin, SV40, and c-*myc* sequences were linked. All 18 transgenic lines developed a disease that was very similar to adult polycystic kidney disease, and died of renal failure within 6 weeks to 3 months of birth.[415] Thus, deregulated c-*myc* expression is both necessary and sufficient to initiate polycystic kidney disease.

Regulation

Unilateral nephrectomy induces a compensatory hypertrophy in the remaining kidney. This was accompanied by a small increase in the expression of c-*myc* RNA in both mice and rabbits that was not much different from the small increase in c-*myc* expression induced by a sham operation.[412,416,417] This contrasts with the situation in liver in which partial hepatectomy causes a big increase in *myc* expression (see Section III.I). However, partial hepatectomy causes increased proliferation, not just hypertrophy. Folic acid is a toxic agent which causes the death of kidney cells and regenerative cell proliferation. This was shown to be accompanied by an increase in the expression of c-*myc* RNA, up 25-fold at 12–18 hours in mice, and up 6-fold at 4 hours in rabbits.[412,417] The increase in the abundance of c-*myc* RNA in folic acid damaged mouse kidney was due to posttranscriptional events.[418] Growth hormone also induced an increase in the level of c-*myc* RNA in rat kidney.[419]

H. Bladder

Bioassay

The potential biological activity of the c-*myc* gene on bladder epithelial cells has been demonstrated in a reconstituted organ experiment.[420] When mouse bladder urothelium was removed, infected with a recombinant retrovirus containing a v-*myc* gene, and placed into the renal capsule of a syngeneic mouse, the cells produced a hyperplasia relative to control cells infected with the vector virus. When a similar experiment was done using a v-*src* oncogene, hyperplasia and dysplasia were produced. However, when a vector containing both *myc* and *src* oncogenes were used focal tumors were produced.[420] The expression of the introduced genes could not be measured in the nonmalignant tissue to determine if *myc* and *src* together were sufficient to get a tumor, or if further events were needed.

Clinical Association

Although we know that activation of the c-*myc* gene is potentially a neoplastic event in bladder, we do not yet have a complete view of how often it is involved, nor do we know a great deal about how it is regulated in bladder urothelium. The c-*myc* gene was not amplified in four stage-II bladder carcinomas.[286] The 3′ half of the c-*myc* gene was found to have a lower level of DNA methylation in tumors, an alteration which was more prominent in tumors of increasing stage.[421] Lower levels of DNA methylation are associated with increased expression, but expression was not measured in the same series of tumors.[421] A high level of c-*myc* RNA was found in hyperplastic and transitional cell carcinoma cell lines generated from carcinogen treated rats.[422] The level of c-*myc* RNA was also elevated in several human bladder cell lines: T24, Hu609, HCV29T, and Hu609T.[423,424] Interestingly, the urothelial cell line HCV29 has much less c-*myc* RNA than its tumorigenic derivative HCV29T.[423] The enhanced expression of the c-*myc* gene in the Hu609 and T24 cell lines was shown to be due to a *trans*-effect, and not to amplification, rearrangement, or mutation.[424,425]

I. Liver

Clinical Associations

Several groups have reported higher levels of c-*myc* RNA in most human liver carcinomas than in nonmalignant human liver.[426–431] One group found the opposite result of less c-*myc* RNA in the carcinomas.[432] The level of c-*myc* RNA in normal liver, but not in carcinomas, has been found to increase during surgery.[433] Thus, the circumstances under which the control liver specimen was collected will have a big impact on interpretation of the results in this type of experiment. The expression of c-Myc protein, measured by Western blot or immunohistochemistry, was also found elevated in hepatocellular carcinomas.[429,431,434,435] Elevated expression of the c-*myc* gene was associated with several preneoplastic states like chronic active hepatitis B infection,[435] and cirrhosis.[428,430,431,434,435]

The level of c-*myc* RNA was found to be very high in the tumorigenic human hepatoma cell line Hep G2, in which *myc* expression was constitutive and not dependent on the growth state of the cells.[436,437] However, these cells did not have amplification or rearrangement of the encoding gene, leaving open the issue of what caused the apparent deregulation.[437]

Hypomethylation is often associated with gene expression. Two groups have found evidence for a relative hypomethylation of the c-*myc* gene in hepatocellular carcinoma compared to nonmalignant liver.[438,439] A potential methylation site in exon 3 was found methylated in normal liver and hypomethylated in 88% of the tumors.[439] Neither group determined whether increased expression of the c-*myc* gene accompanied the hypomethylation. The methylation experiments would have revealed gene amplification, but it was not seen in any of the data shown, nor was such a finding pointed out in the articles.[438,439] In view of the frequent finding of amplification of the *myc* gene in rodent liver cancer (see below), this point needs further study in humans.

It seems there may be more known about the structure and expression of the c-*myc* gene in liver cancer in rats than in humans. Morris hepatomas are rat liver tumors induced by feeding the hepatocarcinogen *N*-(fluoren-2-yl)phthalamic acid, and maintained by intramuscular transplantation in syngeneic rats.[440] Seven Morris hepatomas were studied and found to have an elevated level c-*myc* RNA compared to normal rat liver.[440,441] One of these tumors was found to have a 5- to 10-fold amplification of the c-*myc* gene.[440] Carcinogenesis with *N*-methyl-*N*-nitrosourea resulted in tumors, some of which had elevated c-*myc* expression, and amplification or rearrangement of the gene.[442] The rearrangements were mapped to the 5′ untranslated part of the first exon or just upstream.[442] Seven rat tumors induced with the hepatocarcinogen aflatoxin B_1 all had a variable, but elevated, expression of c-*myc* RNA, with one tumor having a 2- to 4-fold amplification of the encoding gene.[443] Another tumor in this group had a level of c-*myc* gene expression similar to the one with amplification, but lacked both amplification and rearrangement.[443] In another interesting model system, hepatocellular carcinomas which develop spontaneously about 4 months after birth in the LEC strain of rats, after an apparent nonviral hepatitis, were found to have 7- to 30-fold increased expression of c-*myc* RNA.[444] There was no correlation of c-*myc* RNA level with mitotic activity. In fact, the preneoplastic hepatitis displayed a fairly high mitotic index and did not have elevated expression of c-*myc* RNA when both parameters were compared with livers from the parental strain (LEA), which did not have hepatitis or a tendency to liver cancer.[444]

Several systems of hepatocarcinogenesis in rats have been worked out in which a number of preneoplastic states have been defined. These systems are a valuable resource for determining the molecular events in the development of liver cancer. One of the issues addressed in these studies concerns at which stage and cell type is there an alteration in *myc* expression. When hepatocellular carcinoma was induced by feeding a choline-deficient diet, all 13 induced tumors had increased expression and amplification of the c-*myc* gene.[445] In some animals c-*myc* amplification was also seen in nonmalignant portions of the liver. Thus, in this system, amplification of the c-*myc* gene may precede the last stages of tumorigenesis.[445] A conflict is found in the literature on the issue of increased c-*myc* expression preceding tumorigenesis with 3′-methyl-4-dimethylaminoazobenzene: One group found that c-*myc* expression was high in the tumors, but not in the nontumorous parts of the liver,[446] while another found elevated c-*myc* expression in the entire liver during treatment with this carcinogen.[447]

Carcinogenesis with diethylnitrosamine did not cause elevated c-*myc* expression in the whole liver or even in purified γ-glutamyl transpeptidase-positive preneo-plastic cells, but some adenomatous nodules and most of the hepatocellular carcinomas did have elevated expression.[442,448,449] The Solt–Farber carcinogenesis protocol, which involves the carcinogens diethylnitrosamine and 2-acetyl-aminofluorene, and partial hepatectomy for tumor promotion, resulted in increased c-*myc* expression preceding malignancy.[450,451] This was detected in oval cells and basophilic foci by *in situ* hybridization,[450] and in preneoplastic cells purified from the liver by the criteria of a lack of adherence to plates coated with asialofetuin.[451] In both cases the cells that had increased c-*myc* expression corresponded to the cell type from which the carcinomas are believed to arise.[450,451] The Solt–Farber protocol is usually done with male rats because of the faster carcinogenesis, and higher yield of tumors and preneoplastic nodules than in female rats. A corresponding gender difference was detected in the early response of the expression of c-*myc* RNA in the whole liver and in nodules removed at 8 months after initiation; expression was higher in the males.[452–454] The gender difference in c-*myc* expression and carcinogenesis has been associated with differences in growth hormone levels in male and female rats.[452,454] However, even in the females the carcinomas still had elevated expression of c-*myc* RNA.[453,454]

Liver carcinogenesis resulting from a choline-deficient diet, plus the carcinogen ethionine was also accompanied by increased expression of c-*myc* RNA in the preneoplastic oval cells.[455,456] One of the unanswered questions in most of these studies is the significance of elevated expression of the c-*myc* gene in carcinogen-treated livers. Since it is known that elevated expression of c-*myc* is found in cells reentering a proliferative state after a period of quiescence, which will be discussed below for regenerating liver, how can we distinguish elevated expression due to a proliferative stimulus from the hepatocarcinogens from an intrinsic cellular alteration that contributes to transformation. One approach to this problem made use of an oval cell culture prepared from livers of animals getting the choline-deficient ethionine diet for 6 weeks.[456] These cells, when made quiescent by serum deprivation, responded to a proliferative stimulus from serum by transiently increasing their expression of c-*myc* RNA. However, a tumorigenic subline expressed high levels of c-*myc* RNA constitutively.[456] Thus, the early changes in expression of the c-*myc* gene during carcinogenesis may reflect the mitogenic effects of the carcinogenic protocol, but the later alterations in expression are probably due to an intrinsic alteration in regulation of gene expression.

Spontaneous transformation of rat liver epithelial cells in culture can be induced by holding the cells for long periods at confluence.[457,458] Tumorigenic cell lines prepared in this fashion usually had elevated c-*myc* expression compared with sister cell lines that did not get transformed.[458] However, some exceptions have been noted.[457,458]

Infection with hepatitis B virus is one of the most significant risk factors for hepatocellular carcinoma in humans.[459] A direct connection between activation of the c-*myc* oncogene and hepatitis B viruses has been found in several systems. In a survey of 9 hepatocellular carcinomas from woodchucks infected with woodchuck hepatitis virus (WHV), three had a rearrangement and overexpression of the c-*myc* gene.[460] In two of these tumors there was an insertion of a piece of the viral DNA, which contained the viral enhancer, in the vicinity of the c-*myc* gene; one in the 3' untranslated region of exon 3, and the other 600 bp upstream of exon 1.[461] These viral insertions, which are reminiscent of the ALV promoter insertions described earlier, resulted in an increase in expression of the gene.[460,461] The c-*myc* rearrangement in the other tumor resulted in the discovery of the *hcr* gene, which was highly expressed exclusively in normal liver.[462,463] The rearrangement involved the insertion of the *hcr* promoter and part of the *hcr* coding region into intron 1 of the c-*myc* gene, which caused a 50-fold increase in expression, relative to the expression of c-*myc* in normal liver, of an *hcr*-*myc* fusion gene.[462,463] It is unknown if the WHV had a direct role in causing the juxtaposition of the c-*myc* and *hcr* genes, or if the rearrangement was an independent event. Hepatocellular carcinoma in WHV infected animals has also been associated with elevated expression of the c-*myc* gene due to DNA amplification, and also elevated expression of one of the two N-*myc* genes in woodchuck due to viral integration.[464,465] Ground squirrel hepatitis virus is a much weaker carcinogen than WHV, but is associated frequently with elevated expression of the c-*myc* gene due to gene amplification in hepatocellular carcinomas from infected woodchucks[465] and ground squirrels.[466]

In humans, the hepatitis B virus has not been found integrated near the c-*myc* gene. However, evidence has been presented which suggests that the human hepatitis B virus pX protein can *trans*-activate the transcription of the c-*myc* gene.[467] More work clearly needs to be done to determine if there is a connection between the *myc* family and carcinogenesis by the human hepatitis B virus.

Bioassay

In a transgenic mouse experiment, a c-*myc* gene, linked to an albumin enhancer/promoter to direct expression to the liver, caused hepatic dysplasia in young mice and focal adenomas in some mice over 15 months old.[468] No malignancies were noted in the *myc* transgenics. However, when transgenic mice with more than one oncogenic transgene were made, the c-*myc* transgene had a significant carcinogenic effect. For example, SV40 T-antigen transgenic mice got liver tumors in 3–5 months, but with the addition of a c-*myc* transgene they got cancer in 3–6 weeks.[468] Similarly, using recombinant retrovirus vectors to insert oncogenes into a rat liver epithelial cell line, a v-*myc* retrovirus did not cause transformation, but infection with a v-*raf*/v-*myc* combination transformed 2- to 3-fold more efficiently than v-*raf* alone, and caused a hepatocellular carcinoma-like tumor when the cells were injected into nude mice or syngeneic rats, as opposed to the sarcoma-like malignancies caused by v-*raf* alone.[469] In this case, not only did *myc* increase the yield of transformants, but it changed the histology of the resulting tumors.[469] The *myc* does not seem to be as potent an oncogene in liver as SV40 T-antigen or *raf*, but it still has profound biological effects.[468,469]

The insertion of a construct that increases the expression of the c-*myc* gene product into rat liver epithelial cells resulted in a potentiation of cell response to the epidermal growth factor (EGF).[470,471] There was no change in the receptor or the dose response, just an increase in the amount of DNA synthesis that occurred in response to the added growth factor. Neither group noted the appearance of transformed foci.[470,471] When the *myc* construct was inserted with a selectable marker so that the cells which took up DNA could be isolated, a population of cells were prepared all of which presumably had elevated c-*myc* expression.[472] These cells were not morphologically transformed, but did grow to a higher cell density and produced 2.5-fold more colonies in a 14-day colony-forming assay than the parental cells. EGF enhanced the colony-forming ability of the *myc* transfectants, which is consistent with the above experiments.[472]

In another set of experiments, 16 subclones of chemically transformed rat liver epithelial cells were prepared and evaluated for tumorigenicity and gene expression.[473] Tumorigenicity correlated most strongly with elevated expression of the c-*myc* gene. Interestingly, among the eight sublines with the highest expression of c-*myc* RNA, there was a correlation of tumorigenicity with increased expression of transforming growth factor α (TGF-α).[473] This growth factor is homologous with EGF and uses the same receptor. The increased response of cells with exogenous c-*myc* expression to EGF, described in the previous paragraph, has an echo in this tumorigenicity study.[473] Another example of the synergism between EGF/TGF-α and c-*myc* was found in a transgenic mouse experiment in which a *myc* transgene potentiated the activity of a TGF-α transgene.[474] The initial effects of the expression of these genes, seen at 3 weeks of age, included dysplasia and apoptosis. By 16 weeks, over 70% of the mice carried liver nodules of which 25% were hepatocellular carcinomas. Transgenic mice which expressed only the TGF-α transgene took 10–15 months to get liver tumors.[474]

Elevated c-*myc* expression is not always associated with the development of liver tumors. Two spontaneously immortalized rat liver epithelial cell lines had higher expression of the c-*myc* gene than their mortal parental cell type, but tumorigenic derivatives of these cell lines made by transfection of a mutationally activated *ras* gene (N-*ras* or Ha-*ras*), or chemical transformation using aflatoxin B1, did not cause any additional increase in c-*myc* expression.[475] This work suggests an immortalizing function for elevated expression of the c-*myc* gene.[475]

Regulation

Partial hepatectomy results in a regenerative regrowth of the liver. One of the early events in liver regeneration in both mice and rats was an increase in the expression of the c-*myc* gene.[476–484] In mice the increase was seen within the first hour and peaked between 1 and 6 hours at 10- to 100-fold over quiescent liver.[476–478] In rats, the increase in the expression of c-*myc* RNA was also early and transient.[480,483,484] In both species DNA synthesis did not begin for several hours after the early peak in c-*myc* expression. The partial hepatectomy systems confirm the association of an early increase in c-*myc* expression preceding the reentry of a quiescent cell into S phase, as described earlier for fibroblasts. Some groups have found an additional later increase in c-*myc* expression in rats after partial hepatectomy, which corresponded to around the beginning of DNA synthesis.[483,484]

There are several other liver systems in which increased c-*myc* expression is an early event preceding proliferation. Insulin is a mitogen for the rat hepatoma cell lines Reuber H35 and H4IIE, in which it also caused a rapid, but transient, increase in the expression of c-*myc* RNA.[485,486] Epidermal growth factor and transforming growth factor-α also caused an increase in expression of the c-*myc* gene in freshly isolated rat hepatocytes several hours before DNA synthesis started.[471,487–489] The effect of these two factors was rapid (less than 2 hours) and mediated through a pathway involving prostaglandins. Indomethacin, an inhibitor of prostaglandin synthesis, inhibited the TGF-α and EGF effects on *myc* expression and DNA synthesis, but addition of prostaglandins E_2 and $F_{2\alpha}$ restored these effects.[487,489] When the hepatocytes were isolated from older rats, EGF still had the same stimulatory effect on the expression of c-*myc* RNA, but the DNA synthetic response was much less.[488] This last experiment brings up the possibility that increasing c-*myc* expression does not force proliferation in hepatocytes.

In a number of other experimental systems elevated expression of the c-*myc* gene did not correlate with increased cellular proliferation in liver cells. A sham operation itself, without removing any liver tissues, caused an increase in c-*myc* expression in the liver of mice, rats, and humans, although the magnitude was usually less and the timing different from the response to partial hepatectomy.[433,478,484] Inflammation caused by the intraperitoneal injection of Freund's adjuvant led to an increase in the level of c-*myc* RNA in rat liver.[484] Another example of a lack of correspondence between increased c-*myc* expression and proliferation was found when freshly isolated hepatocytes were put into culture; there was an initial transient surge in expression of c-*myc* RNA, then a more sustained increase which was not followed by cell proliferation.[490,491] Similarly, intraperitoneal injection of glycine in rats caused a rapid transient increase in c-*myc* expression which was not followed by increased DNA synthesis.[492] Deprivation of protein caused an increase in the expression of c-*myc* RNA slowly over several days, which was rapidly reversed by refeeding a protein rich diet.[493] DNA synthesis followed several hours after c-*myc* RNA levels fell to normal.[493] Thus, increased expression of the c-*myc* gene does not necessarily lead to proliferation. We speculate that in some of these systems Myc may be performing a function related to apoptosis. Earlier we noted an association between expression of c-*myc* and TGF-α transgenes, and apoptosis in the liver.[474]

The growth inhibitor and differentiation agent butyrate caused a decrease in expression of the c-*myc* gene in several rat hepatoma cell lines.[494,495] In the HTC line the effect of butyrate was very rapid regardless of the cell cycle phase the cells were in, and was not blocked by the protein synthesis inhibitor cycloheximide which indicates a direct effect of the agent that is not mediated by the synthesis of a new protein(s).[494]

Postpubertal male rats had a twofold increased level of c-*myc* RNA in their livers compared with females.[496] When the male rats were castrated, c-*myc* expression equalized between the sexes. The gender differences were reestablished if the castrated rats were given testosterone. When intact males were given additional growth hormone by osmotic minipump for 1 week, their expression of c-*myc* RNA was decreased to the female level.[496] The evidence suggests that gender differences in the expression of growth hormone, which are caused by testosterone, may account for gender differences in the effectiveness of some carcinogenesis protocols (described above), and that the regulation of expression of the c-*myc* gene in the liver by growth hormone may be part of the mechanism.[452–454,496] However, another group found that growth hormone had the opposite effect of rapidly and transiently raising the level of expression of the c-*myc* gene in the livers of hypophysectomized rats *in vivo*.[419] These differences are more apparent than real since the experiments involved differences in the dose of growth hormone in both the controls and the experimental animals, strain of rats, and timing of growth hormone administration and harvesting of the liver for analysis.[419,496]

J. Pancreas

Clinical Associations

An early report on one adenocarcinoma of the pancreas showed a low level of c-*myc* RNA,[407] while another group reported a pancreatic carcinoma with c-*myc* amplification and elevated expression of the gene.[497] In adenomas and adenocarcinomas of the pancreas, induced in rats by azaserine, the expression of c-*myc* RNA was elevated 2- to 8-fold and 8- to 40-fold, respectively.[498] The level of *myc* RNA in the azaserine induced carcinomas was found to be greater than in regenerating nonmalignant rat pancreas, although both tissues had similar rates of proliferation.[498] Four cases of endocrine pancreatic cancer have been reported as having a level of c-*myc* RNA below that of normal pancreas.[499] Pancreatic cancer has been relatively neglected by those studying the structure and expression of the *myc* gene in clinical specimens.

Bioassay

The biological impact on the pancreas of elevated expression of the c-*myc* gene has been studied in a number of ways. The beta cells of the fetal rat pancreas were stimulated to proliferate by electroporation of a construct that expressed the human c-*myc* gene under control of an insulin promoter, and the effect was more consistent when an activated Ha-*ras* gene was cotransfected.[500] However, continuously growing lines were not established.[500] The low level of expression in cancer of the endocrine cells, described above, and the lack of major biological activity of the c-*myc* gene in these cells, leads to the suspicion that this gene may not play a role in the rare cancers of the endocrine pancreas.

Infection of newborn mice with a recombinant retrovirus containing the v-*myc* oncogene caused acinar carcinomas in 31% with a latency of 150 days.[501] The latency and the fact that the tumors were clonal suggests that additional events were required to produce neoplasia.[501]

Early experiments in which transgenic mice expressed c-*myc* in the pancreas failed to develop neoplasia in that organ.[314,502] But when the transgene was expressed at a high enough level by using an elastase promoter/enhancer to stimulate transcription and direct it to the pancreas, and a growth hormone 3′ untranslated region and polyadenylation site with the intent of stabilizing the usually very unstable c-*myc* RNA,[40] carcinomas of the pancreas were induced between 2 and 7 months of age.[183] The initial effect of the c-*myc* transgene was to produce a dysplasia in the acinar cells, but the carcinomas were of both acinar and mixed acinar ductal histology. It seems that the ductal carcinomas arose from transformed acinar cells. The *myc* transgene influenced the histology of the resultant tumors, as similar transgenic mice with activated *ras* or SV40 T-antigen transgenes developed only acinar adenocarcinomas.[183,502] The transgenic mice experiments implicate the c-*myc* gene as a potential element in the pathogenesis of ductal adenocarcinoma of the pancreas, which is the most common type of pancreatic cancer in man.[183] Clearly, overexpression of the *myc* oncogene can be a factor in pancreatic carcinogenesis; however the relative dearth of clinical studies does not allow a judgment about how frequently.

Regulation

Regeneration of the rat pancreas in response to partial pancreatectomy was preceded by an increase in expression of c-*myc* RNA.[503] *myc* RNA was up more than twofold at 12 hours after surgery, which was before the peak of DNA synthesis seen at 2 days.[503] It will be interesting to look at earlier events in regeneration of the pancreas in view of the rapid changes in c-*myc* expression seen during liver regeneration (see Section I. on liver). Camostat, a low molecular weight proteinase inhibitor, has been shown to cause the secretion of cholecystokinin which then results in a wave of DNA synthesis in rat pancreas *in vivo*; a transient 14-fold increase in the level of c-*myc* RNA preceded DNA synthesis.[498]

Several agents which stimulate secretion by pancreatic acinar cells, cholecystokinin, bombesin, and carbachol caused a small increase in the level of c-*myc* RNA in freshly isolated rat acinar cells, while other agents which stimulate secretion such as gastrin, secretin, or vasoactive intestinal peptide had no effect.[504] Thus, elevated c-*myc* expression is not necessary for secretion in this tissue.

K. Stomach and Esophagus

Clinical Associations

Elevated expression of the c-*myc* oncogene has been shown to be rare in carcinoma of the esophagus.[505-507] However, some tumors with increased expression of c-*myc* RNA or protein have been found.[505-507]

The expression of protein and RNA encoded by the c-*myc* gene was found elevated compared to nonmalignant gastric mucosa in about half of the primary stomach cancers that have been reported.[505,508] One group found that expression of c-Myc protein in the stromal cells of gastric carcinomas correlated with lower invasiveness and a better prognosis.[509] In this study, staining of the tumor cells was not predictive of stage, metastatic ability, or prognosis.[509] In inflammatory, metaplastic, and dysplastic lesions, increased expression of c-Myc protein was found in some cases.[508,510,511] However, an extremely high level of c-Myc protein in nonmalignant specimens was found only in some adenomatous polyps and dysplastic lesions.[511] Amplification of the c-*myc* gene was found to be more rare, occurring in 3 out of the 52 primary stomach carcinomas that have been reported,[286,512,513] and in 3 of 16 xenografts maintained in nude mice.[514] In a study that compared clinical specimens of gastric and colorectal carcinoma, the bowel malignancies had greater levels of c-*myc* RNA.[505] The biological significance of increased expression of the c-*myc* gene in gastric and esophageal cancer is unknown at this time.

Regulation

In vitro, serum starved AGS gastric carcinoma cells responded to serum with an increase in proliferation and expression of c-*myc* RNA, effects which were diminished by simultaneous exposure to both vasoactive intestinal polypeptide and isobutylmethyl-xanthine.[515] These agents increase cAMP production and implicate this signaling pathway in the regulation of c-*myc* expression and growth in gastric epithelial cells.[515] The gastric carcinoma cell line TMK-1 responded to the mitogen epidermal growth factor with a transient four-fold increase in expression of c-*myc* RNA at 1 hour, which preceded the increase in DNA synthesis.[516]

L. Colon and Rectum

Clinical Associations

In a survey of 29 primary human adenocarcinomas of the large bowel, 72% had a 5- to 40-fold elevated level of c-*myc* RNA compared with nonmalignant mucosa.[517] Additional surveys have revealed similar frequencies of elevated expression of c-*myc* RNA.[407,505,518-526] Surveys of adenomatous polyps, which are considered premalignant lesions, showed about two-thirds had a modestly elevated level of c-*myc* RNA that was lower than in the carcinomas.[525,526] Therefore, overexpression of c-*myc* can occur early in the neoplastic process.

Questions have been raised as to whether overexpression of the c-*myc* gene is a significant event in carcinogenesis or simply a marker of increased cellular proliferation. In one study enhanced expression of the c-*myc* gene in colorectal adenocarcinoma was paralleled by an increase in the expression of the cell-cycle genes 2A9, ornithine decarboxylase (ODC), and histone H3, compared to normal mucosa in five of the six tumors studied.[527] If the expression of the c-*myc* gene was the rate-limiting step in cellular proliferation, then we would expect such a result. However, it has been suggested that the increased expression of the c-*myc* gene in these tumors reflects their greater proliferative capacity, and that a "true" overexpression of the c-*myc* gene would result in an increase in the relative abundance of c-*myc* RNA compared to the expression of cell-cycle-dependent genes like the histones.[527]

At this point, an examination of the kinetics of cellular proliferation in normal and malignant colonic mucosa might prove illuminating. Turnover of normal mucosa is faster than the potential doubling time of tumor cells assuming no cell loss occurs.[528–530] The S phase of the cell cycle is much longer in carcinoma cells than in normal cells.[531] At any given time more cells are in S phase in tumor than in normal mucosa, and this probably accounts for the tumor-specific increase in expression of the histone genes, which are expressed only during S phase.[532] Thus, histone gene expression is probably not a good tool to evaluate the proliferation rate in tumors of the large bowel. The expression of the c-*myc* gene is not confined to the S phase of the cell cycle.[533–535] The growth of colon tumors appears to be due to a lack of maturation and loss of the mature cells in the lumen of the gut, which is the fate of normal colonocytes.

Interestingly, even using the stringent criteria for overexpression of the c-*myc* gene that it has to be out of proportion to the expression of histone RNA, 56% of 25 tumors examined in one study fulfilled this criteria.[536] Other experiments have also confirmed a lack of correspondence between markers of cellular proliferation and expression of the c-*myc* gene. In a study of azoxymethane-induced colon carcinogenesis in F344 male rats, the expression of c-*myc* RNA was elevated in all of the adenomas and invasive carcinomas studied, compared to slight and variable increases for Ha-*ras*, ODC, and β-actin.[537] In an immunohistochemical study, there was no relationship between Myc expression in human colorectal carcinomas and cellular proliferation; the latter being evaluated using the Ki-67 antibody.[538] Therefore, the level of expression of the c-*myc* gene does not seem to be an indicator of the rate of cellular proliferation in tumors of the large bowel.

Using immunohistochemistry, a number of laboratories have studied the expression of Myc protein in normal and malignant colonic mucosa. This methodology allows the determination of the percentage and localization of cells expressing Myc. In normal colonic mucosa, embedded in paraffin, immunohistochemistry with the monoclonal antibody (Mab) 6E10 revealed maximal staining in the cytoplasm of cells in the maturation zone between the proliferating cells of the basal crypt and the mature surface epithelium.[539,540] However, staining of frozen tissue with several different anti-Myc antibodies differed with respect to the intracellular location and the location within the crypt: Myc was largely limited to the nuclei of cells in the basal crypt.[538,541] One study using paraffin-embedded tissue demonstrated cytoplasmic and nuclear staining in 40–50% of the cells in the lower half of the crypt using the Mab 6E10, compared with nuclear staining in 25–30% of cells in the lower third of the crypt using the Mabs H51C116 and H8C150.[538] Explanations for these discrepancies have been suggested: (1) nuclear localization of Myc may be lost after fixation,[538,539,542,543] and/or (2) cytoplasmic staining with 6E10 may be due to cross-reactivity of the Mab.[540,541] *In situ* hybridization has been used to show that c-*myc* RNA is located in the base of the crypt.[519] Thus it is likely that Myc protein is also located in that region. Most recent studies agree that Myc is located in the nucleus of normal colonic mucosa.[538,541]

Analysis of adenomatous polyps by immunohistochemistry revealed a broader distribution of positive staining cells extending from the basal crypts to the luminal surface.[538,539,544] In adenocarcinomas of the large bowel the intensity and number of staining cells increased even more dramatically.[538–541,545] Thus, Myc protein is expressed at an elevated level in this malignancy consistent with the results from RNA studies. However, there is a great deal of heterogeneity in the staining patterns between tumors. Some adenocarcinomas expressed Myc in every cell, in contrast with others that had mixtures of both positively and negatively stained regions.[538,540] Two groups that found a nuclear location for Myc in normal mucosa found evidence for a cytoplasmic location in carcinomas.[538,543] This difference in localization may be due either to an artifact of fixation or it may suggest that Myc is less tightly associated with the nuclear matrix in tumor cells.[538,543] However, colon carcinoma cell lines stained exclusively in the nucleus.[538,546]

Interestingly, histologically normal mucosa immediately adjacent to some adenocarcinomas gave a positive staining pattern identical to the tumors.[538] This phenomenon may be due to a preneoplastic change, or since it was not observed in adenomas, a factor secreted by adenocarcinomas that causes elevated expression of the c-*myc* gene in tumor tissue as well as in the adjacent normal mucosa.[538]

Although elevated expression of the c-*myc* gene occurs in over 70% of primary colorectal carcinomas, gross amplification or rearrangement of the gene is pretty rare. Amplification of the c-*myc* gene was detected in only 4.7% of 232 patient samples analyzed.[517,520–522,524–526,547–549] Very slight—less than twofold—amplification of the c-*myc* gene was found more frequently and was associated with rare, aggressive subtypes of colon cancer.[550] In a survey of 13 mucinous and 7 poorly differentiated tumors, 50% revealed a low level of c-*myc* amplification, compared to less than 7% for 29 moderately and well differentiated tumors. The amplification in these tumors (with one exception) was not due to polysomy of the chromosome carrying the c-*myc* gene, since the c-*mos* gene, also located on chromosome 8,[19] was not similarly amplified. Slight amplification may be due to a subpopulation of cells within the tumor with a substantial amplification of the c-myc gene.[550]

Amplification of the c-myc gene has been demonstrated in 7 (COLO320, HT29, SW480, SW620, SW742, WiDr, NCI-H716) of 29 colon carcinoma cell lines analyzed.[255,517,522,551-555] Amplification of the c-myc gene was also seen in a mouse colon cancer induced by dimethylhydrazine.[556] Cell lines and tumors with significant amplification of the c-myc gene had elevated levels of c-myc RNA,[520,522,524-526,546,553-556] with one exception.[525] These levels of c-myc expression were similar to tumors and cell lines without amplification of the c-myc gene.[522,546,553,555] Therefore, elevated expression of c-myc can occur with and without gene amplification.

Rearrangement of the c-myc gene in colon cancer has only been reported in two cell lines, COLO320DM[557] and SW480,[425,546] and one patient specimen.[547] In each case the rearrangement was associated with amplification of the gene, which may have contributed to the rearrangement by providing additional targets. The c-myc gene rearrangement in COLO320DM was not found in its sister cell line COLO320HSR, suggesting that the rearrangement may have taken place after establishment of these cells in culture.[554] Although gross rearrangements of the c-myc gene are rare, the possibility of more subtle genetic alterations cannot be ruled out. In endemic Burkitt's lymphoma, mutations in the exon 1/intron 1 boundary region of the gene have been implicated in causing an increase in the expression of c-myc RNA.[558,559] DNA sequence analysis of this region in a group of colorectal carcinoma cell lines with elevated expression of c-myc RNA (HCT8, NCI-H498, NCI-H747, SW837, SW1116, WiDr) revealed no alterations in the sequence.[560] Thus, an endemic Burkitt's lymphoma-like mutational event is unlikely to be responsible for the deregulation of c-myc expression seen in colorectal carcinoma.

Methylation of DNA is a covalent alteration which is often correlated with decreased gene expression.[561] DNA from normal colonic mucosa possessed a high level of methylation of a site in the third exon of the c-myc gene, while adenomatous polyps and adenocarcinomas had significantly less methylation at this site.[562] Expression of the c-myc gene was not analyzed in this study so we do not know if the hypomethylated state had the expected impact on gene expression.[562] The rectal carcinoma cell line SW837 had both elevated expression of the c-myc gene and significant hypomethylation of the exon 3 site, while the WiDr cell line had elevated c-myc expression, but was completely methylated at the same site.[553] However, the WiDr cell line had an amplification of the c-myc gene which was probably responsible for the elevated level of c-myc RNA in these cells. The expression of c-myc RNA per gene copy was essentially normal in the WiDr cells, which is consistent with its normal methylation state. The hypomethylation of the c-myc gene in SW837, which did not have c-myc amplification, is consistent with demethylation being contributory to elevated c-myc expression.[553]

The potential clinical value of knowing about an abnormal structure and/or expression of the c-*myc* gene in a tumor is of great interest. Earlier we discussed the association of slight amplification with aggressive subtypes of colorectal cancer. On the expression side, several studies, including a total of 162 tumors, have not found a correlation of elevated c-*myc* expression with a number of clinical parameters, such as the histological type, stage, depth of invasion, recurrence of disease, or the survival rate.[518,523–525,551] However, an interesting correlation has been found between elevated expression of c-*myc* RNA and the location of the tumor within the colon.[12,563] Elevated expression of the c-*myc* gene occurs more frequently in tumors of the left side (distal) than in tumors of the right side (proximal). It has been argued that increased *myc* expression may be a marker for a genetically distinct form of colon cancer that is the sporadic (noninherited) version of the colorectal tumors that frequently occur in patients with the inherited disease adenomatous polyposis coli.[12,13,563–565]

What causes the elevated expression of c-*myc* in those colorectal tumors that do not have amplification or rearrangement of the gene? The lack of any evidence for a *cis* alteration suggests that a *trans* mechanism is at work. In support of this, the overexpression of the c-*myc* gene and the tumorigenic phenotype were suppressed by fusion of colon carcinoma cells with cells that express low levels of c-*myc* RNA.[564,565] Based on the association of elevated expression with adenocarcinomas of the distal large bowel, the *APC* gene (adenomatous polyposis coli locus) was suggested as a potential regulator of the expression of the c-*myc* gene in colorectal mucosa.[12,13,563] This hypothesis is supported by the finding that elevated c-*myc* expression in colorectal adenocarcinoma correlated with a loss of heterozygosity for markers on human chromosome 5q,[564] which is the location of the *APC* tumor suppressor gene.[566] Furthermore, transfer of a normal chromosome 5 into the DLD-1, SW1116, and COKFu colorectal carcinoma cell lines suppressed the overexpression of c-*myc*, as well as tumorigenicity in nude mice.[565,567] Abnormal regulation of expression of the c-*myc* gene may, therefore, be one of the consequences of loss of function of the *APC* gene.

Two experiments have provided evidence that an activated *ras* gene may be responsible for increasing the expression of the c-*myc* gene in colon cancer. In the first experiment, mutationally activated human Ha-*ras* gene, under the control of a MMTV promoter, was inserted into rat intestinal cells.[568] Expression of this construct, induced by dexamethasone, resulted in the elevated expression of c-*myc* RNA and an immortalized, but nontumorigenic, phenotype.[568] The second experiment involved the disruption of either the normal or mutated endogenous Ki-*ras* gene in DLD-1 and HCT116 colon carcinoma cells by homologous recombination.[569] While inactivation of the normal allele had no effect on either cell line, inactivation of the mutationally activated Ki-*ras* gene resulted in an altered morphology, a slowing of cell growth, an inhibition of growth in soft agar, a loss of tumorigenicity in nude mice, and a 10-fold decrease in expression of the c-*myc* gene in both lines.[569] Although there is evidence for both *ras* and *APC* mutations being responsible for causing elevated c-*myc* expression in colon cancer, the detailed mechanisms, and relative importance of *ras* activation, *APC* inactivation, and other yet to be discovered events are unknown.

Bioassay

The clinical studies provide evidence that c-*myc* activation may be an important step in producing neoplasia of the large bowel. In support of this, a cause and effect relationship between elevated expression of the c-*myc* gene and the proliferative potential of COLO320 cells has been established.[169] Treatment of COLO320 cells with a 15-base antisense oligonucleotide covering the translation initiation site located in exon 2 of the c-*myc* gene resulted in a 40–75% decrease in the ability of the cells to form colonies in soft agar, which was dependent on the concentration of the antisense oligonucleotide. Oligonucleotides with a variant sequence, either sense or missense, were not effective.[169]

The biological consequences of exogenous expression of the *myc* gene and other oncogenes in rat colon has been studied using retroviral vectors.[570–572] Fragments of rat fetal colon tissue, maintained in collagen gels, were infected with retroviral vectors containing combinations of the v-*myc*, v-*src* or v-Ha-*ras* genes under control of strong promoters and evaluated for growth in culture.[571] Infection with a single oncogene resulted in a greater outgrowth of epithelial cells, which underwent senescence and death within 2 weeks just like uninfected fragments, but did not result in immortalized cell lines. Tissue infected with a combination of either v-*myc* and v-Ha-*ras*, or v-*myc* and v-*src* produced immortalized colonic epithelial cell lines in about 80% of the attempts. Cells transformed with v-*myc* and Ha-*ras* displayed a classical epithelial morphology, sucrase isomaltase activity, and the expression of keratin filaments. Cells transformed with v-*myc* and v-*src* were less adherent, more mesenchymal in morphology, possessed sucrase isomaltase activity, and did not express keratin filaments.[571] In a similar series of experiments, segments of descending colon and rectum, dissected from rat fetuses, were infected with retroviral vectors containing v-*myc*, v-*src*, or v-*myc* and v-*src* together.[572] Infected colonic segments were transplanted subcutaneously into syngeneic rats and the established heterotopic implants were harvested after 60–100 days. Out of 16 transplants infected with a v-*myc* retrovirus, 5 had focal areas of atypia or dysplasia and 1 displayed goblet cell hyperplasia. All 12 of the v-*src*-infected transplants had focal areas of atypia/dysplasia with one sarcoma. The combination of v-*myc* and v-*src* produced one adenocarcinoma and 11 transplants with atypia and high-grade dysplasia out of 16 successful transplants. The combination of v-*myc* and v-*src* resulted in a more severely dysplastic histology than v-*myc* alone, and a more precarcinomatous histology than the metaplasia produced by v-*src*.[572] These experiments provide evidence for significant biological activity of the *myc* gene in the colorectum. However, since tumorigenicity was not the endpoint of most of the experiments, we do not have as clear a picture as we would like of the potential role of the c-*myc* gene in carcinogenesis in this tissue.

Regulation

Treatment of colon carcinoma cells with sodium butyrate may provide a model system for the early molecular events involved in differentiation of the epithelial cells of the colonic mucosa. Butyrate treatment of colon cancer cell lines caused altered morphology, a slowing of cell growth, and an inhibition of growth in soft agar.[169,573–577] The differentiation markers carcinoembryonic antigen and a placental-type alkaline phosphatase were also induced by butyrate.[147,560,573,574,576,578–580] A decrease in the expression of c-*myc* RNA has also been associated with exposure to sodium butyrate in several colon carcinoma cell lines.[147,169,577,581,582] Treatment of the rectal carcinoma cell line SW837 with 2 mM of sodium butyrate caused a rapid decline in the expression of c-*myc* RNA that was blocked by protein synthesis inhibitors, suggesting that butyrate caused the synthesis of a protein which has a negative effect on the abundance of c-*myc* RNA.[147] Treatment of SW837 cells with 2 mM butyrate for 6 hours caused a decrease in transcriptional elongation through the c-*myc* gene of sufficient magnitude to account for the 10-fold decrease in the level of c-*myc* RNA expression.[560] This effect was accompanied by a change in promoter usage.[560] Therefore, the protein(s) induced by butyrate may inactivate or block the activity of a factor involved in transcriptional elongation, or it may have a direct role in limiting elongation. Conversely, butyrate induced a decrease in the abundance of c-*myc* RNA due to posttranscriptional alterations, CACO-2 cells.[582] Thus, there may be heterogeneity in the mechanism by which butyrate lowers c-*myc* expression.

Agents other than butyrate have also been used to investigate the regulation of the c-*myc* gene. Treatment of COLO320 cells with 2-difluoromethylornithine, a suicide inhibitor of ornithine decarboxylase, depleted intracellular polyamines resulting in an inhibition of cell growth and a 90% decrease in the expression of c-*myc* RNA that was due to a decrease in the transcription rate of the c-*myc* gene.[583,584] These results suggest that polyamine metabolism is involved in regulating the transcription of the c-*myc* gene. Transcription of the c-*myc* gene was also inhibited (42% at 96 hours) in DLD-1 Clone A cells treated with *N*-methylformamide.[149] This agent also caused a decrease in DNA synthesis, cell growth, and tumorigenicity in this human colon cancer cell line.[149] Exposure of DLD-1 Clone A cells to recombinant human interferon-β ser[17] (IFN-β ser[17]) caused a dose dependent reduction in expression of the c-*myc* gene and an inhibition of cellular proliferation.[585] In this case, though, the stability of c-*myc* RNA was altered, with its half-life decreasing from 29 minutes to 15 minutes after a 4-day treatment with IFN-β ser[17]. The evidence suggested that the activity of the 2′,5′ oligoadenylate synthetase/ RNase L pathway, which is activated by IFN-β, may destabilize c-*myc* RNA.[585]

Treatment of the human colorectal carcinoma cell lines HCT116 and MOSER with the differentiation agent N,N-dimethylformamide (DMF) resulted in a reduction of c-*myc* expression, which did not require synthesis of new protein.[148] Proliferating MOSER cells also responded to the transforming growth factor β (TGF-β₁) with a decrease in the expression of c-*myc* RNA.[148,586] The reduction of c-*myc* expression upon exposure to DMF or TGF-β₁ was associated with a decrease in cellular proliferation and a more benign phenotype.[148]

Tumors are different from person to person, and even within an individual tumor there is heterogeneity. A tissue culture model system may be useful in understanding a tumor cell type, but an understanding of cancer must involve an appreciation of tumor cell heterogeneity. Colon carcinoma cell lines representing different tumor cell types have been developed and characterized.[587] These cell lines have been classified into groups based upon their state of differentiation *in vitro*.[587] Poorly differentiated cells (group I) grow in a monolayer and form poorly differentiated xenografts in nude mice. Well-differentiated cells (group III), which resemble normal epithelial cells, grow in three-dimensional patches and display brush-borden microvilli and tight junctions. Compared to group III cells, group I cells are more tumorigenic *in vivo* and have higher plating efficiencies in soft agar.

A series of investigations has explored differences in the way these cells respond to their environment with respect to growth and regulation of gene expression. In growing cell cultures, the expression of the c-*myc* gene did not differ appreciably between cell lines of the well- and poorly differentiated groups.[588] However, they did differ in the way the c-*myc* gene was regulated. Quiescent group III cells responded to the replenishment of nutrients and the growth factors epidermal growth factor, insulin, and transferrin (EIT) with an increase in both expression of c-*myc* RNA and DNA synthesis: When nutrients and EIT were added to quiescent cultures of group III cell lines, FET or CBS, the expression of c-*myc* RNA increased fivefold after 4 hours of treatment and was back down to base line by 24 hours; DNA synthesis peaked at around 20 hours.[589,590] Both responses were inhibited by TGF-β₁.[589,591] Interestingly, quiescent group I cells did not respond to EIT or TGF-β₁, but did increase DNA synthesis when nutrients were replenished.[590] Group I HCT116 cells did not respond to EIT with increased DNA synthesis or increased expression of c-*myc* RNA.[590,591] When nutrients were replenished in the quiescent HCT116 cell cultures, DNA synthesis increased but the expression of c-*myc* RNA did not increase.[590,591] Clearly, there is a big difference in the way these two different cell types respond to growth stimulation. They differ as well in response to growth inhibition using TGF-β or deprivation of growth factors. TGF-β₁ caused a decrease in expression of c-*myc* RNA in the group III cells but not in the group I cells.[591] The group I cell lines, HCT116 and RKO, increased their expression of c-*myc* RNA when deprived of growth factors, while the group III lines, FET and CBS, did not change appreciably.[588] The group III cells usually follow the classic pattern of responding to growth stimulation by increasing expression of the c-*myc* gene, and growth inhibition by decreasing expression. The group I cells respond in a seemingly inappropriate way or do not respond at all. An important lesson from these studies is the hazards of generalizing from results generated with one or a few cell lines. Not only is there tissue-specific differences in the regulation of the c-*myc* gene, but even in malignancies of the same tissue there is heterogeneity. The molecular mechanisms behind the differences in control of c-*myc* expression and the impact of these alterations on the other phenotypic differences will tell an interesting story.

Animal models for the modulation of c-*myc* expression have also been established. All-*trans* retinoic acid significantly reduced the total number of aberrant crypt foci induced by azoxymethane (AOM) in Sprague–Dawley rats.[592] The number of aberrant crypts expressing c-*myc* RNA decreased from 82% in the AOM-treated rats to 25% in the rats treated with AOM and retinoic acid.[592] While retinoic acid decreases c-*myc* expression, estradiol may act to increase expression of the c-*myc* gene. Tumors, resulting from injection of MC-26 mouse colon carcinoma cells into Balb/c mice, were consistently larger (by weight) in females, suggesting that female gonadal hormones play a role in tumor growth.[593] Ovariectomized (OVX) mice injected with MC-26 cells presented significantly smaller tumors after 21 days than OVX mice treated with estradiol or sham-operated control mice. Increased doses of estradiol produced larger tumors in both male and female mice injected with MC-26 cells. The larger tumors induced by estrogen treatment had higher levels of expression of ODC and c-*myc* RNA.[593] A direct effect of estrogen on expression of the c-*myc* gene was not established in this study, as in estrogen-responsive breast cancer cell lines (see Section III.B).

Radiation causes an increase in proliferation of cells in the colonic crypts in rats and mice, which compensates for the cells lost to the toxic insult.[594,595] This increase in proliferation was accompanied by an increase in expression of the c-*myc* gene. The Bowman–Birk protease inhibitor (BBI), which has anticarcinogenic properties in several systems, inhibited the increase in expression of c-*myc* RNA seen in X-irradiated mouse colon and in γ-irradiated rat colon at a 7-day timepoint. Interestingly, BBI had no effect on the proliferation of the crypt cells, or on the expression of c-*myc* RNA in unirradiated colon.[594,595] These results suggest that a protease may be involved in the pathway for elevating c-*myc* expression.[594] Since the anticarcinogenic property of the BBI was not tested in this system, we do not know if the inhibition of expression of the c-*myc* gene is associated with a reduction in carcinogenesis.

IV. PUTTING THINGS IN PERSPECTIVE: DOES *MYC* CAUSE CANCER EVERYWHERE?

It is clear from the vast volume of work on the biological activity of the c-*myc* gene that it is a potentially carcinogenic agent in many tissues. One exception to this trend is the salivary gland. In the MMTV-*myc* transgenic mouse strains developed by the Leder laboratory, the salivary gland was one of the tissues with the highest levels of *myc* expression, but not a single tumor was noted in this tissue.[312,314,315] MMTV-*ras* transgenic mice got salivary gland tumors, so there was no barrier to a MMTV-oncogene construct causing cancer in this organ.[315] The salivary gland appears resistant to the carcinogenic effects of elevated expression of the c-*myc* gene. The barrier may be the absence of some target or cofactor required for c-*myc* carcinogenicity.

Cancer has long been considered the product of a number of independent events.[596–598] The evaluation of the biological potency of an activated c-*myc* gene in the systems described in this review reveal that in every case c-*myc* could not induce cancer by itself. Whenever it was evaluated, the tumors were monoclonal or oligoclonal proliferations in a tissue that had an activated c-*myc* gene in the nonmalignant cells. Activation of c-*myc* expression increased the probability of a cell becoming a tumor, but did not assure that result. Even in experiments in which an activated *myc* gene was combined with another activated oncogene, the tumors were clonal. *Thus, activation of the c-myc oncogene is not sufficient for carcinogenesis.*

Another finding from the study of transgenic animals is that activated c-*myc* expression in many tissues is compatible with grossly normal development.[89–91,312–314,385,415,468,474,599] Because of this, abnormal regulation of expression of the c-*myc* gene may be considered a possible facet in cancer predisposition. As described in the section on liver, there was a correlation between gender differences in regulation of the c-*myc* gene and susceptibility to hepatocellular carcinoma in a strain of rats. Polymorphisms in the regulation of *myc* expression may turn out to be significant in understanding genetic differences in susceptibility to malignancy.

Not all cellular proliferation is associated with increased expression of the c-myc gene. Examples include the C3H10T1/2 mouse fibroblast cell line in which the proteinase inhibitor antipain caused a reduction in the expression of c-*myc* RNA, and a disappearance of the transient increase in expression which was seen when quiescent cells were stimulated to proliferate, without altering DNA synthesis, rate of proliferation, or saturation density.[600,601] Similar findings have been made *in vivo*: During early development of the human placenta the highest level of c-*myc* expression corresponded to the cytotrophoblastic shell which is very proliferative,[602] however in the fetus there was a low level of c-*myc* RNA despite rapid proliferation at 3–4 weeks postconception.[603] At 6–10 weeks postconception highly proliferative epithelial layers had a high level of expression of c-*myc* RNA, but the cartilage cells of the rapidly growing limb bud and head had a low level of c-*myc* RNA.[603] In mouse embryos the expression of c-*myc* RNA decreased sharply at 7.5 days in the most proliferative cells, the primitive ectoderm.[604] Later in development of the mouse embryo there was a good correspondence between proliferation rate and c-*myc* expression.[605,606] As described earlier, adult mouse spermatogonia actively proliferate and have very low levels of c-*myc* RNA.[401] The association between increased expression of the c-*myc* gene and cellular proliferation does not hold in many normal tissues, so elevated *myc* expression is not simply a marker of proliferation.

Does carcinogenesis require activation of c-*myc*? As described earlier, the c-*myc* gene is overexpressed very frequently in colorectal cancer especially when it occurs in the distal colon, but not every adenocarcinoma of the large bowel has overexpression. On the other end of the spectrum, elevated c-*myc* expression was found much less frequently in leukemia, especially chronic lymphocytic leukemia in which elevated expression was rarely found.[607-609] As seen throughout this review, there are many examples of malignancy without elevated expression of the c-*myc* gene. *Cancer does not require elevated c-myc expression.*

Even though a c-*myc* function appears to be a requirement for carcinogenesis in many experimental systems, there probably are ways to circumvent the need for overexpression of the c-*myc* gene itself. An example of this kind of carcinogenic complementation is found in the study of p53 mutations and mdm2 amplification[610]: The mdm2 gene product binds to and inhibits the tumor suppressor p53. Amplification and overexpression of mdm2 bypasses the need for a p53 mutation in carcinogenesis. Similarly, in some types of cancer an activated N-*myc* gene seems to substitute for c-*myc* activation.[111,112,611,612] However, we do not yet have a clear picture of exactly what might qualify as a substitute for c-*myc* activation as a required step in carcinogenesis.

Despite the extensive literature on the c-*myc* gene we still lack an understanding of how this gene causes cancer. However, a continuation of the rapid progress in the last few years on understanding the biochemical activities of Myc protein makes it likely that its role will become gradually clearer. The genes that are regulated by c-Myc and the rules that govern when the quantity of Myc protein is the critical variable in their regulation will be an important focus of future work. The events that result in activation of the c-*myc* gene are known for some cancers, and it is likely that additional mechanisms will be clear in the near future, perhaps involving the inactivation of tumor suppressor genes. There is still a mountain of work to be done before the *myc* gene is truly understood.

ACKNOWLEDGMENTS

This work was supported by National Cancer Institute Grant CA50246 to P.G.R., and a Postdoctoral Fellowship to D.P.H. awarded by the Scientific Education Partnership, which is funded through the Marion Merrell Dow Foundation. The laboratory was also supported by a grant from the Hall Family Foundations of Kansas City and the Patton Memorial Trust. We thank Darren Baker for help in gathering references. Despite our attempt to be comprehensive it is likely that we have omitted important references through oversight. For this and any unintentional misinterpretations we ask forgiveness.

REFERENCES

1. Marcu, K. B.; Bossone, S. A.; Patet, A. J. *myc* function and regulation. *Annu. Rev. Biochem.* **1992,** *61,* 809–860.
2. Erisman, M. D.; Astrin, S. M. The *myc* oncogene. In: *The Oncogene Handbook* (Reddy, E. P.; Skalka, A. M.; Curran T., Eds.) Elsevier Science Publishers B.V., New York, 1988, pp. 341–379.
3. DePinho, R. A.; Schreiber-Argus, N.; Alt, F. W. *myc* family oncogenes in the development of normal and neoplastic cells. *Adv. Cancer Res.* **1991,** *57,* 1–46.
4. Spencer, C. A.; Groudine, M. Control of c-*myc* regulation in normal and neoplastic cells. *Adv. Cancer Res.* **1991,** *56,* 1–48.
5. Murphy, L. J. Estrogen induction of insulin-like growth factors and *myc* protooncogene expression in the uterus. *J. Steroid Biochem. Molec. Biol.* **1991,** *40,* 223–230.
6. Neckers, L. M.; Rosolen, A.; Whitesell, L. Antisense inhibition of gene expression. In: *Gene Regulation: Biology of Antisense RNA and DNA* (Erickson, R.P.; Izant, J.G., Eds.) Raven Press, New York, 1992, pp. 295–302.
7. Pillai, R. Oncogene expression and prognosis in cervical cancer. *Cancer Lett.* **1991,** *59,* 171–175.
8. Luscher, B.; Eisenman, R. N. New light on *myc* and *myb.* Part I. Myc. *Genes Dev.* **1990,** *4,* 2025–2035.
9. Field, J. K.; Spandidos, D. A. The role of *ras* and *myc* oncogenes in human solid tumors and their relevance in diagnosis and prognosis (review). *Anticancer Res.* **1990,** *10,* 1–22.
10. Riou, G. F. Proto-oncogenes and prognosis in early carcinoma of the uterine cervix. *Cancer Surv.* **1988,** *7,* 441–456.
11. Frauman, A. G.; Moses, A. C. Oncogenes and growth factors in thyroid carcinogenesis. *Endocrinol. & Metabol. Clin. North America* **1990,** *19,* 479–493.
12. Rothberg, P. G. The role of the oncogene c-*myc* in sporadic large bowel cancer and familial polyposis coli. *Semin. Surg. Oncol.* **1987,** *3,* 152–158.
13. Astrin, S. M.; Costanzi, C. The molecular genetics of colon cancer. *Semin. Oncol.* **1989,** *16,* 138–147.
14. Shiu, R. P. C.; Watson, P. H.; Dubik, D. c-*myc* oncogene expression in estrogen-dependent and -independent breast cancer. *Clin. Chem.* **1993,** *39,* 353–355.
15. Dalla-Favera, R. Chromosomal translocations involving the c-*myc* oncogene and their role in the pathogenesis of B-cell neoplasia. In: *Origins of Human Cancer* (Brugge, J.; Curran, T.; Harlow, E.; McCormick, F., Eds.) Cold Spring Harbor Laboratory Press, Cold Spring Harbor, NY, 1991, pp. 534–551.
16. Schrier, P. I.; Peltenburg, L. T. C. Relationship between *myc* oncogene activation and MHC class I expression. *Adv. Cancer Res.* **1993,** *60,* 181–246.
17. Dalla-Favera, R.; Bregni, M.; Erikson, J.; Patterson, D.; Gallo, R. C.; Croce, C. M. Human c-*myc onc* gene is located on the region of chromosome 8 that is translocated in Burkitt lymphoma cells. *Proc. Natl. Acad. Sci. USA* **1982,** *79,* 7824–7827.
18. Taub, R.; Kirsch, I.; Morton, C.; et al. Translocation of the c-*myc* gene into the immunoglobulin heavy-chain locus in human Burkitt lymphoma and murine plasmacytoma cells. *Proc. Natl. Acad. Sci. USA* **1982,** *79,* 7837–7841.
19. Neel, B. G.; Jhanwar, S. C.; Chaganti, R. S.; Hayward, W. S. Two human c-*onc* genes are located on the long arm of chromosome 8. *Proc. Natl. Acad. Sci. USA* **1982,** *79,* 7842–7846.
20. Gazin, C.; Dupont de Dinechin, S.; Hampe, A.; et al. Nucleotide sequence of the human c-*myc* locus: provocative open reading frame within the first exon. *EMBO J.* **1984,** *3,* 383–387.
21. Battey, J.; Moulding, C.; Taub, R.; et al. The human c-*myc* oncogene: structural consequences of translocation into the IgH locus in Burkitt Lymphoma. *Cell* **1983,** *34,* 779–787.
22. Colby, W. W.; Chen, E. Y.; Smith, D. H.; Levinson, A. D. Identification and nucleotide sequence of a human locus homologous to the v-*myc* oncogene of avian myelocytomatosis virus MC29. *Nature* **1983,** *301,* 722–725.

23. Hayday, A. C.; Gillies, S. D.; Saito, H.; et al. Activation of a translocated human c-*myc* gene by an enhancer in the immunoglobulin heavy-chain locus. *Nature* **1984**, *307*, 334–340.

24. Eick, D.; Polack, A.; Kofler, E.; Lenoir, G. M.; Rickinson, A. B.; Bornkamm, G. W. Expression of P$_0$- and P$_3$-RNA from the normal and translocated c-*myc* allele in Burkitt's lymphoma cells. *Oncogene* **1990**, *5*, 1397–1402.

25. Bentley, D. L.; Groudine, M. Novel promoter upstream of the human c-*myc* gene and regulation of c-*myc* expression in B-cell lymphomas. *Mol. Cell. Biol.* **1986**, *6*, 3481–3489.

26. Hann, S. R.; King, M. W.; Bentley, D. L.; Anderson, C. W.; Eisenman, R. N. A non-AUG translational initiation in c-*myc* exon I generates an N-terminally distinct protein whose synthesis is disrupted in Burkitt's lymphomas. *Cell* **1988**, *52*, 185–195.

27. Perrson, H.; Hennighausen, L.; Taub, R.; DeGrado, W.; Leder, P. Antibodies to human c-*myc* oncogene product: evidence of an evolutionarily conserved protein induced during cell proliferation. *Science* **1984**, *225*, 687–693.

28. Hann, S. R.; Eisenman, R. N. Proteins encoded by the human c-*myc* oncogene: Differential expression in neoplastic cells. *Mol. Cell. Biol.* **1984**, *4*, 2486–2497.

29. Ramsay, G.; Evan, G. I.; Bishop, J. M. The protein encoded by the human proto-oncogene c-*myc*. *Proc. Natl. Acad. Sci. USA* **1984**, *81*, 7742–7746.

30. Dedieu, J-F.; Gazin, C.; Rigolet, M.; Galibert, F. Evolutionary conservation of the product of human c-*myc* exon I and its inducible expression in a murine cell line. *Oncogene* **1988**, *3*, 523–529.

31. Gazin, C.; Rigolet, M.; Briand, J. P.; Van Regenmortel, M. H. V.; Galibert, F. Immunochemical detection of proteins related to the human c-*myc* exon 1. *EMBO J.* **1986**, *5*, 2241–2250.

32. Eladari, M. E.; Syed, S. H.; Guilhot, S.; d'Auriol, L.; Galibert, F. On the high conservation of the human c-*myc* first exon. *Biochem. Biophys. Res. Commun.* **1986**, *140*, 313–319.

33. Persson, H.; Leder, P. Nuclear localization and DNA binding properties of a protein expressed by human c-*myc* oncogene. *Science* **1984**, *225*, 718–721.

34. Winqvist, R.; Saksela, K.; Alitalo, K. The *myc* proteins are not associated with chromatin in mitotic cells. *EMBO J.* **1984**, *3*, 2947–2950.

35. Evan, G. I.; Hancock, D. C. Studies on the interaction of the human c-*myc* protein with cell nuclei: p62$^{c\text{-}myc}$ as a member of a discrete subset of nuclear proteins. *Cell* **1985**, *43*, 253–261.

36. Spector, D. L.; Watt, R. A.; Sullivan, N. F. The v- and c-*myc* oncogene proteins colocalize in situ with small nuclear ribonucleoprotein particles. *Oncogene* **1987**, *1*, 5–12.

37. Luscher, B.; Eisenman, R. N. Mitosis-specific phosphorylation of the nuclear oncoproteins *myc* and *myb*. *J. Cell Biol.* **1992**, *118*, 775–784.

38. Brewer, G.; Ross, J. Poly(A) shortening and degradation of the 3' A + U-rich sequences of human c-*myc* mRNA in a cell-free system. *Mol. Cell. Biol.* **1988**, *8*, 1697–1708.

39. Jones, T. R.; Cole, M. D. Rapid cytoplasmic turnover of c-*myc* cRNA: requirement of the 3' untranslated sequences. *Mol. Cell. Biol.* **1987**, *7*, 4513–4521.

40. Dani, C.; Blanchard, J. M.; Piechaczyk, M.; El Sabouty, S.; Marty, L.; Jeanteur, P. Extreme instability of *myc* mRNA in normal and transformed human cells. *Proc. Natl. Acad. Sci. USA* **1984**, *81*, 7046–7050.

41. Rabbitts, P. H.; Forster, A.; Stinson, M. A.; Rabbitts, T. H. Truncation of exon 1 from the c-*myc* gene results in prolonged c-*myc* mRNA stability. *EMBO J.* **1985**, *4*, 3727–3733.

42. Roussel, M.; Saule, S.; Lagrou, C.; et al. Three new types of viral oncogene of cellular origin specific for haematopoietic cell transformation. *Nature* **1979**, *281*, 452–455.

43. Graf, T.; Beug, H. Avian leukemia viruses: interaction with their target cells *in vitro* and *in vitro*. *Biochim. Biophys. Acta* **1978**, *516*, 269–299.

44. Alitalo, K.; Bishop, J. M.; Smith, D. H.; Chen, E. Y.; Colby, W. W.; Levinson, A. D. Nucleotide sequence of the v-*myc* oncogene of avian retrovirus MC29. *Proc. Natl. Acad. Sci. USA* **1983**, *80*, 100–104.

45. Reddy, E. P.; Reynolds, R. K.; Watson, D. K.; Schultz, R. A.; Lautenberger, J.; Papas, T. S. Nucleotide sequence analysis of the proviral genome of avian myelocytomatosis virus (MC29). *Proc. Natl. Acad. Sci. USA* **1983**, *80*, 2500–2504.

46. Bister, K.; Hayman, M. J.; Vogt, P. K. Defectiveness of avian myelocytomatosis virus MC29: isolation of long-term nonproducer cultures and analysis of virus-specific polypeptide synthesis. *Virol.* **1977**, *82*, 431–448.

47. Coll, J.; Righi, M.; de Taisne, C.; Dissous, C.; Gegonne, A.; Stehelin, D. Molecular cloning of the avian acute transforming retrovirus MH2 reveals a novel cell-derived sequence (v-*mil*) in addition to the *myc* oncogene. *EMBO J.* **1983**, *2*, 2189–2194.

48. Hayflick, J.; Seeburg, P. H.; Ohlsson, R.; et al. Nucleotide sequence of two overlapping myc-related genes in avian carcinoma virus OK10 and their relation to the *myc* genes of other viruses and the cell. *Proc. Natl. Acad. Sci. USA* **1985**, *82*, 2718–2722.

49. Kan, N. C.; Flordellis, C. S.; Garon, C. F.; Duesberg, P. H.; Papas, T. S. Avian carcinoma virus MH2 contains a transformation-specific sequence, *mht*, and shares the *myc* sequence with MC29, CMII and OK10 viruses. *Proc. Natl. Acad. Sci. USA* **1983**, *80*, 6566–6570.

50. Saule, S.; Sergeant, A.; Torpier, G.; Raes, M. B.; Pfeifer, S.; Stehelin, D. Subgenomic mRNA in OK10 defective leukemia virus-transformed cells. *J. Virol.* **1982**, *42*, 71–82.

51. Pachl, C.; Biegalke, B.; Linial, M. RNA and protein encoded by MH2 virus: evidence for subgenomic expression of v-*myc*. *J. Virol.*. **1983**, *45*, 133–139.

52. Saule, S.; Coll, J.; Righi, M.; Lagrou, C.; Raes, M. B.; Stehelin, D. Two different types of transcription for the myelocytomatosis viruses MH2 and CMII. *EMBO J.* **1983**, *2*, 805–809.

53. Dogget, D. L.; Drake, A. L.; Hirsch, V.; Rowe, M. E.; Stallard, V.; Mullins, J. I. Structure, origin, and transforming activity of feline leukemia virus-*myc*-recombinant provirus FTT. *J. Virol.* **1989**, *63*, 2108–2117.

54. Neil, J. C.; Hughes, D.; McFarlane, R.; et al. Transduction and rearrangement of the *myc* gene by feline leukaemia virus in naturally occurring T-cell leukaemias. *Nature* **1984**, *308*, 814–820.

55. Levy, L. S.; Gardner, M. B.; Casey, J. W. Isolation of a feline leukaemia provirus containing the oncogene *myc* from a feline lymphosarcoma. *Nature* **1984**, *308*, 853–856.

56. Mullins, J. I.; Brody, D. S.; Binari, R. C., Jr.; Cotter, S. M. Viral transduction of c-*myc* gene in naturally occurring feline leukemias. *Nature* **1984**, 856–858.

57. Fulton, R.; Forrest, D.; McFarlane, R.; Onions, D.; Neil, J. C. Retroviral transduction of T-cell antigen receptor β-chain and *myc* genes. *Nature* **1987**, *326*, 190–194.

58. Ramsay, G.; Graf, T.; Hayman, M. J. Mutants of avian myelocytomatosis virus with smaller *gag* gene-related proteins have an altered transforming ability. *Nature* **1980**, *288*, 170–172.

59. Enrietto, P. J.; Hayman, M. J.; Ramsay, G. M.; Wyke, J. A.; Payne, L. N. Altered pathogenicity of avian myelocytomatosis (MC29) viruses with mutations in the v-*myc* gene. *Virol.* **1983**, *124*, 164–172.

60. Bister, K.; Ramsay, G. M.; Hayman, M. J. Deletions within the transformation-specific RNA sequences of acute leukemia virus MC29 give rise to partially transformation-defective mutants. *J. Virol.* **1982**, *41*, 754–766.

61. Ramsay, G. M.; Hayman, M. J. Isolation and biochemical characterization of partially transformation-defective mutants of avian myelocytomatosis virus strain MC29: localization of the mutation to the *myc* domain of the 110,000-dalton *gag-myc* polyprotein. *J. Virol.* **1982**, *41*, 745–753.

62. Enrietto, P. J. A small deletion in the carboxy terminus of the viral *myc* gene renders the virus MC29 partially transformation defective in avian fibroblasts. *Virol.* **1989**, *168*, 256–266.

63. Zhou, R-P.; Kan, N.; Papas, T.; Duesberg, P. Mutagenesis of avian carcinoma virus MH2: Only one of two potential transforming genes (δ*gag-myc*) transforms fibroblasts. *Proc. Natl. Acad. Sci. USA* **1985**, *82*, 6389–6393.

64. Vennstrom, B.; Sheiness, D.; Zabielski, J.; Bishop, J. M. Isolation and characterization of c-*myc*, a cellular homolog of the oncogene (v-*myc*) of avian myelocytomatosis virus strain 29. *J. Virol.* **1982**, *42*, 773–779.

65. Robins, T.; Bister, K.; Garon, C.; Papas, T.; Duesberg, P. Structural relationship between a normal chicken DNA locus and the transforming gene of the avian acute leukemia virus MC29. *J. Virol.* **1982**, *41*, 635–642.

66. Jansen, H. W.; Patschinsky, T.; Bister, K. Avian oncovirus MH2: molecular cloning of proviral DNA and structural analysis of viral RNA and protein. *J. Virol.* **1983**, *48*, 61–73.

67. Walther, N.; Jansen, H. W.; Trachmann, C.; Bister, K. Nucleotide sequence of CMII v-*myc* allele. *Virol.* **1986**, *154*, 219–223.

68. Ramsay, G. M.; Moscovici, G.; Moscovici, C.; Bishop, J. M. Neoplastic transformation and tumorigenesis by the human protooncogene MYC. *Proc. Natl. Acad. Sci. USA* **1990**, *87*, 2102–2106.

69. Filardo, E. J.; Humphries, E. H. An avian retrovirus expressing chicken $pp55^{c-myc}$ possesses weak transforming activity distinct from v-*myc* that may be modulated by adjacent normal cell neighbors. *J. Virol.* **1991**, *65*, 6621–6629.

70. Symonds, G.; Hartshorn, A.; Kennewell, A.; O'Mara, M-A.; Bruskin, A.; Bishop, J. M. Transformation of murine myelomonocytic cells by *myc*: point mutations in v-*myc* contribute synergistically to transforming potential. *Oncogene* **1989**, *4*, 285–294.

71. Frykberg, L.; Graf, T.; Vennstrom, B. The transforming activity of the chicken c-*myc* gene can be potentiated by mutations. *Oncogene* **1987**, *1*, 415–421.

72. Linial, M. Two retroviruses with similar transforming genes exhibit differences in transforming potential. *Virol.* **1982**, *119*, 382–391.

73. Royer-Pokora, B.; Beug, H.; Claviez, M.; Winkhardt, H. J.; Friis, R. R.; Graf, T. Transformation parameters in chicken fibroblasts transformed by AEV and MC29 avian leukemia viruses. *Cell* **1978**, *13*, 751–760.

74. Palmieri, S.; Kahn, P.; Graf, T. Quail embryo fibroblasts transformed by four v-*myc*-containing virus isolates show enhanced proliferation but are not tumorigenic. *EMBO J.* **1983**, *2*, 2385–2389.

75. Vennstrom, B.; Kahn, P.; Adkins, B.; et al. Transformation of mammalian fibroblasts and macrophages *in vitro* by a murine retrovirus encoding an avian v-*myc* oncogene. *EMBO J.* **1984**, *3*, 3223–3229.

76. Davis, A. C.; Wims, M.; Spotts, G. D.; Hann, S. R.; Bradley, A. A null c-*myc* mutation causes lethality before 10.5 days of gestation in homozygotes and reduced fertility in heterozygous female mice. *Genes Dev.* **1993**, *7*, 671–682.

77. Hayward, W. S.; Neel, B. G.; Astrin, S. M. Activation of a cellular onc gene by promoter insertion in ALV-induced lymphoid leukosis. *Nature* **1981**, *290*, 475–480.

78. Payne, G. S.; Bishop, J. M.; Varmus, H. E. Multiple arrangements of viral DNA and an activated host oncogene in bursal lymphomas. *Nature* **1982**, *295*, 209–214.

79. Noori-Daloii, M. R.; Swift, R. A.; Kung, H-J.; Crittenden, L. B.; Witter, R. L. Specific integration of REV proviruses in avian bursal lymphomas. *Nature* **1981**, *294*, 574–576.

80. Li, Y.; Holland, C. A.; Hartley, J. W.; Hopkins, N. Viral integration near c-*myc* in 10–20% of MCF 247-induced AKR lymphomas. *Proc. Natl. Acad. Sci. USA* **1984**, *81*, 6808–6811.

81. Selten, G.; Cuypers, H. T.; Zijlstra, M.; Melief, C.; Berns, A. Involvement of c-*myc* in MuLV-induced T cell lymphomas in mice: frequency and mechanisms of activation. *EMBO J.* **1984**, *3*, 3215–3222.

82. Corcoran, L. M.; Adams, J. M.; Dunn, A. R.; Cory, S. Murine T lymphomas in which the cellular *myc* oncogene has been activated by retroviral insertion. *Cell* **1984**, *37*, 113–122.

83. Makowski, D. R.; Rothberg, P. G.; Astrin, S. M. The role of promoter insertion in the induction of neoplasia. *Surv. Synth. Pathol. Res.* **1984**, *3*, 342–349.

84. Makowski, D. R.; Rothberg, P. G.; Astrin, S. M. Cellular transformation by avian viruses. *Pharmac. Ther.* **1985**, *27*, 63–97.

85. Shen-Ong, G. L. C.; Keath, E. J.; Piccoli, S. P.; Cole, M. D. Novel *myc* oncogene RNA from abortive immunoglobulin-gene recombination in mouse plasmacytomas. *Cell* **1982**, *31*, 443–452.

86. Marcu, K. B.; Harris, L. J.; Stanton, L. W.; Erikson, J.; Watt, R.; Croce, C. M. Transcriptionally active c-*myc* oncogene is contained within NIARD, a DNA sequence associated with chromosome translocations in B-cell neoplasia. *Proc. Natl. Acad. Sci. USA* **1983**, *80*, 519–523.

87. Hamlyn, P. H.; Rabbitts, T. H. Translocation joins c-*myc* and immunoglobulin τl Genes in a Burkitt Lymphoma revealing a third exon in the c-*myc* oncogene. *Nature* **1983**, *304*, 135–139.

88. Adams, J. M.; Gerondakis, S.; Webb, E.; Corcoran, L. M. S. C. Cellular *myc* oncogene is altered by chromosome translocation to an immunoglobulin locus in murine plasmacytomas and is rearranged similarly in human Burkitt lymphomas. *Proc. Natl. Acad. Sci. USA* **1983**, *80*, 1982–1986.

89. Schmidt, E. V.; Pattengale, P. K.; Weir, L.; Leder, P. Transgenic mice bearing the human c-*myc* gene activated by an immunoglobulin enhancer: A pre-B-cell lymphoma model. *Proc. Natl. Acad. Sci. USA* **1988**, *85*, 6047–6051.

90. Knight, K. L.; Spieker-Polet, H.; Kazdin, D. S.; Oi, V. T. Transgenic rabbits with lymphocytic leukemia induced by the c-*myc* oncogene fused with the immunoglobulin heavy chain enhancer. *Proc. Natl. Acad. Sci. USA* **1988**, *85*, 3130–3134.

91. Adams, J. M.; Harris, A. W.; Pinkert, C. A.; et al. The c-*myc* oncogene driven by immunoglobulin enhancers induces lymphoid malignancy in transgenic mice. *Nature* **1985**, *318*, 533–538.

92. Langdon, W. Y.; Harris, A. W.; Cory, S.; Adams, J. M. The c-*myc* oncogene perturbs B lymphocyte development in Eμ-*myc* transgenic mice. *Cell* **1986**, *47*, 11–18.

93. Baumbach, W. R.; Stanley, E. R.; Cole, M. D. Induction of clonal monocyte-macrophage tumors *in vivo* by a mouse c-*myc* retrovirus: Rearrangement of the CSF-1 gene as a secondary transforming event. *Mol. Cell. Biol.* **1987**, *7*, 664–671.

94. Baumbach, W. R.; Keath, E. J.; Cole, M. D. A mouse c-*myc* retrovirus transforms established fibroblast lines *in vitro* and induces monocyte-macrophage tumors *in vivo*. *J. Virol.* **1986**, *59*, 276–283.

95. Keath, E. J.; Caimi, P. G.; Cole, M. D. Fibroblast lines expressing activated c-*myc* oncogenes are tumorigenic in nude mice and syngeneic animals. *Cell* **1984**, *39*, 339–348.

96. Kelekar, A.; Cole, M. D. Tumorigenicity of fibroblast lines expressing the adenovirus E1a, cellular p53, or normal c-*myc* genes. *Mol. Cell. Biol.* **1986**, *6*, 7–14.

97. Pellegrini, S.; Basilico, C. Rat fibroblasts expressing high levels of human c-*myc* transcripts are anchorage-independent and tumorigenic. *J. Cell. Physiol.* **1986**, *126*, 107–114.

98. Zerlin, M.; Julius, M. A.; Cerni, C.; Marcu, K. B. Elevated expression of an exogenous c-*myc* gene is insufficient for transformation and tumorigenic conversion of established fibroblasts. *Oncogene* **1987**, *1*, 19–27.

99. Land, H.; Parada, L. F.; Weinberg, R. A. Tumorigenic conversion of primary embryo fibroblasts requires at least two cooperating oncogenes. *Nature* **1983**, *304*, 596–602.

100. Land, H.; Chen, A. C.; Morgenstern, J. P.; Parada, L. F.; Weinberg, R. A. Behavior of *myc* and *ras* oncogenes in transformation of rat embryo fibroblasts. *Mol. Cell. Biol.* **1986**, *6*, 1917–1925.

101. Lee, W. M. F.; Schwab, M.; Westaway, D.; Varmus, H. E. Augmented expression of normal c-*myc* is sufficient for cotransformation of rat embryo cells with a mutant *ras* gene. *Mol. Cell. Biol.* **1985**, *5*, 3345–3356.

102. Alt, F. W.; Kellems, R. E.; Bertino, J. R.; Schimke, R. T. Selective multiplication of dihydrofolate reductase genes in methotrexate-resistant variants of cultured murine cells. *J. Biol. Chem.* **1978**, *253*, 1357–1370.

103. Dalla-Favera, R.; Wong-Staal, F.; Gallo, R. C. *Onc* gene amplification in promyelocytic leukaemia cell line HL60 and primary leukaemic cells of the same patient. *Nature* **1982**, *299*, 61–63.

104. Collins, S.; Groudine, M. Amplification of endogenous *myc*-related DNA sequences in a human myeloid leukemia cell line. *Nature* **1982**, *298*, 679–681.

105. Campisi, J.; Gray, H. E.; Pardee, A. B.; Dean, M.; Sonenshein, G. E. Cell-cycle control of c-*myc* but not c-*ras* expression is lost following chemical transformation. *Cell* **1984**, *36*, 241–247.

106. Grignani, F.; Lombardi, L.; Inghirami, G.; Sternas, L.; Cechova, K.; Dalla-Favera, R. Negative autoregulation of c-*myc* expression is inactivated in transformed cells. *EMBO J.* **1990**, *9*, 3913–3922.

107. Martinotti, S.; Richman, A.; Hayday, A. Disruption of the putative c-*myc* auto-regulation mechanism in a human B cell line. *Curr. Topic Microbiol. & Immunol.* **1988**, *141*, 264–268.

108. Kohl, N. E.; Kanda, N.; Schreck, R. R.; et al. Transposition and amplification of oncogene-related sequences in human neuroblastomas. *Cell* **1983**, *35*, 359–367.

109. Schwab, M.; Alitalo, K.; Klempnauer, K-H.; et al. Amplified DNA with limited homology to *myc* cellular oncogene is shared by human neuroblastoma cell lines and a neuroblastoma tumour. *Nature* **1983**, *305*, 245–248.

110. Nau, M. M.; Brooks, B. J.; Battey, J.; et al. L-*myc*, a new *myc*-related gene amplified and expressed in human small cell lung cancer. *Nature* **1985**, *318*, 69–73.

111. Schwab, M.; Varmus, H. E.; Bishop, J. M. Human N-*myc* gene contributes to neoplastic transformation of mammalian cells in culture. *Nature* **1985**, *316*, 160–162.

112. Yancopoulos, G. D.; Nisen, P. D.; Tesfaye, A.; Kohl, N. E.; Goldfarb, M. P.; Alt, F. W. N-*myc* can cooperate with *ras* to transform normal cells in culture. *Proc. Natl. Acad. Sci. USA* **1985**, *82*, 5455–5459.

113. Birrer, M. J.; Segal, S.; DeGreve, J. S.; Kaye, F.; Sausville, E. A.; Minna, J. D. L-*myc* cooperates with *ras* to transform primary rat embryo fibroblasts. *Mol. Cell. Biol.* **1988**, *8*, 2668–2673.

114. Sugiyama, A.; Kume, A.; Nemoto, K.; et al. Isolation and characterization of s-*myc*, a member of the rat *myc* gene family. *Proc. Natl. Acad. Sci. USA* **1989**, *86*, 9144–9148.

115. Ingvarrson, S.; Asker, C.; Axelson, H.; Klein, G.; Sumegi, J. Structure and expression of B-*myc*, a new member of the *myc* gene family. *Mol. Cell. Biol.* **1988**, *8*, 3168–3174.

116. Resar, L. M. S.; Dolde, C.; Barrett, J. F.; Dang, C. V. B-*myc* inhibits neoplastic transformation and transcriptional activation by c-*myc*. *Mol. Cell. Biol.* **1993**, *13*, 1130–1136.

117. Lee, S-Y.; Sugiyama, A.; Sueoka, N.; Kuchino, Y. Point mutation of the *neu* gene in rat neural tumor RT4-AC cells: suppression of tumorigenicity by s-*myc*. *Jpn. J. Cancer Res.* **1990**, *81*, 1085–1088.

118. Suen, T-C.; Hung, M-C. c-*myc* reverses *neu*-induced transformed morphology by transcriptional repression. *Mol. Cell. Biol.* **1991**, *11*, 354–362.

119. Alt, F. W.; DePinho, R.; Zimmerman, K.; et al. The human *myc* gene family. Cold Spring Harbor *Symp. Quant. Biol.* **1986**, *51*, 931–941.

120. Mougneau, E.; Lemieux, L.; Rassoulzadegan, M.; Cuzin, F. Biological activities of v-*myc* and rearranged c-*myc* oncogenes in rat fibroblast cells in culture. *Proc. Natl. Acad. Sci. USA* **1984**, *81*, 5758–5762.

121. Tavassoli, M.; Shall, S. Transcription of the c-*myc* oncogene is altered in spontaneously immortalized rodent fibroblasts. *Oncogene* **1987**, *2*, 337–345.

122. Kelly, K.; Cochran, B. H.; Stiles, C. D.; Leder, P. Cell-specific regulation of the c-*myc* gene by lymphocyte mitogens and platelet-derived growth factor. *Cell* **1983**, *35*, 603–610.

123. Armelin, H. A.; Armelin, M. C. S.; Kelly, K.; et al. Functional role for c-*myc* in mitogenic response to platelet-derived growth factor. *Nature* **1984**, *310*, 655–660.

124. Kaczmarek, L.; Hyland, J. K.; Watt, R.; Rosenberg, M.; Baserga, R. Microinjected c-*myc* as a competence factor. *Science* **1985**, *228*, 1313–1315.

125. Cavalieri, F.; Goldfarb, M. Growth factor-deprived BALB/c 3T3 murine fibroblasts can enter the S phase after induction of c-*myc* gene expression. *Mol. Cell. Biol.* **1987**, *7*, 3554–3560.

126. Stern, D. F.; Roberts, A. B.; Roche, N. S.; Sporn, M. B.; Weinberg, R. A. Differential responsiveness of *myc*- and *ras*-transfected cells to growth factors: Selective stimulation of *myc*-transfected cells by epidermal growth factor. *Mol. Cell. Biol.* **1986**, *6*, 870–877.

127. Sorrentino, V.; Drozdoff, V.; McKinney, M. D.; Zeitz, L.; Fleissner, E. Potentiation of growth factor activity by exogenous c-*myc* expression. *Proc. Natl. Acad. Sci. USA* **1986**, *83*, 8167–8171.

128. Paulsson, Y.; Bywater, M.; Heldin, C-H.; Westermark, B. Effects of epidermal growth factor and platelet-derived growth factor on c-*fos* and c-*myc* mRNA levels in normal human fibroblasts. *Exp. Cell Res.* **1987**, *171*, 186–194.

129. Rozengurt, E.; Sinnett-Smith, J. W. Bombesin induction of c-*fos* and c-*myc* protooncogenes in Swiss 3T3 cells: significance for the mitogenic response. *J. Cell. Physiol.* **1987**, *131*, 218–225.

130. Cutry, A. F.; Kinniburgh, A. J.; Twardzik, D. R.; Wenner, C. E. Transforming growth factor alpha (TGFα) induction of c-*fos* and c-*myc* expression in C3H 10T1/2 cells. *Biochem. Biophys. Res. Commun.* **1988**, *152*, 216–222.

131. Ran, W.; Dean, M.; Levine, R. A.; Henkle, C.; Campisi, J. Induction of c-*fos* and c-*myc* mRNA by epidermal growth factor or calcium ionophore is cAMP dependent. *Proc. Natl. Acad. Sci. USA* **1986**, *83*, 8216–8220.

132. Morrow, M. A.; Lee, G.; Gillis, S.; Yancopoulos, G. D.; Alt, F. W. Interleukin-7 induces N-*myc* and c-*myc* expression in normal precursor B lymphocytes. *Genes Dev.* **1992**, *6*, 61–70.

133. Lacy, J.; Sarkar, S. N.; Summers, W. C. Induction of c-*myc* expression in human B lymphocytes by B-cell growth factor and anti-immunoglobulin. *Proc. Natl. Acad. Sci. USA* **1986**, *83*, 1458–1462.

134. Snow, E. C.; Fetherston, J. D.; Zimmer, S. Induction of the c-*myc* protooncogene after antigen binding to hapten-specific B cells. *J. Exp. Med.* **1986**, *164*, 944–949.

135. Reed, J. C.; Nowell, P. C.; Hoover, R. G. Regulation of c-*myc* mRNA levels in normal human lymphocytes by modulators of cell proliferation. *Proc. Natl. Acad. Sci. USA* **1985**, *82*, 4221–4224.

136. Spangler, R.; Bailey, S. C.; Sytkowski, A. J. Erythropoietin increases c-*myc* mRNA by a protein kinase C-dependent pathway. *J. Biol. Chem.* **1991**, *266*, 681–684.

137. Westin, E. H.; Wong-Staal, F.; Gelmann, E. P.; et al. Expression of the cellular homologs of retroviral onc genes in human hematopoietic cells. *Proc. Natl. Acad. Sci USA* **1982**, *79*, 2490–2494.

138. Reitsma, P. H.; Rothberg, P. G.; Astrin, S. M.; et al. Regulation of *myc* gene expression in HL-60 cells by a vitamin D metabolite. *Nature* **1983**, *306*, 492–494.

139. Gonda, T. J.; Metcalf, D. Expression of *myc*, *myb* and *fos* protooncogenes during the differentiation of a murine myeloid leukaemia. *Nature* **1984**, *310*, 249–251.

140. Larsson, L-G.; Ivhed, I.; Gidlund, M.; Pettersson, U.; Vennstrom, B.; Nilsson, K. Phorbol ester-induced terminal differentiation is inhibited in human U-937 monoblastic cells expressing a v-*myc* oncogene. *Proc. Natl. Acad. Sci. USA* **1988**, *85*, 2638–2642.

141. Tonini, G. P.; Radzioch, D.; Gronberg, A.; et al. Erythroid differentiation and modulation of c-*myc* expression induced by antineoplastic drugs in the human leukemic cell line K562. *Cancer Res.* **1987**, *47*, 4544–4547.

142. Lachman, H. M.; Skoultchi, A. I. Expression of c-*myc* changes during differentiation of mouse erythroleukemia cells. *Nature* **1984**, *310*, 592–594.

143. Endo, T.; Nadal-Ginard, B. Transcriptional and posttranscriptional control of c-*myc* during myogenesis: Its mRNA remains inducible in differentiated cells and does not suppress the differentiated phenotype. *Mol. Cell. Biol.* **1986**, *1986*, 1412–1421.

144. Sejersen, T.; Sumegi, J.; Ringertz, N. R. Density-dependent arrest of DNA replication is accompanied by decreased levels of c-*myc* mRNA in myogenic but not differentiation-defective myoblasts. *J. Cell. Physiol.* **1985**, *125*, 465–470.

145. Griep, A. E.; DeLuca, H. F. Decreased c-*myc* expression is an early event in retinoic acid-induced differentiation of F9 teratocarcinoma cells. *Proc. Natl. Acad. Sci. USA* **1986**, *83*, 5539–5543.

146. de Bustros, A.; Baylin, S. B.; Berger, C. L.; Roos, B. A.; Leong, S. S.; Nelkin, B. D. Phorbol esters increase calcitonin gene transcription and decrease c-*myc* mRNA levels in cultured human medullary thyroid carcinoma. *J. Biol. Chem.* **1985**, *260*, 98–104.

147. Herold, K. M.; Rothberg, P. G. Evidence for a labile intermediate in the butyrate induced reduction of the level of c-*myc* RNA in SW837 rectal carcinoma cells. *Oncogene* **1988**, *3*, 423–428.

148. Mulder, K. M.; Brattain, M. G. Alterations in c-*myc* expression in relation to maturational status of human colon carcinoma cells. *Int. J. Cancer* **1988**, *42*, 64–70.

149. Chatterjee, D.; Mendelsohn, A.; Shank, P. R.; Savarese, T. M. Reversible suppression of c-*myc* expression in a human colon carcinoma line by the anticancer agent N-Methylformamide. *Cancer Res.* **1989**, *49*, 3910–3916.

150. Dotto, G. P.; Gilman, M. Z.; Maruyama, M.; Weinberg, R. A. c-*myc* and c-*fos* expression in differentiating mouse primary keratinocytes. *EMBO J.* **1986**, *5*, 2853–2857.

151. Nath, P.; Getzenberg, R.; Beebe, D.; Pallansch, L.; Zelenka, P. c-*myc* mRNA is elevated as differentiating lens cells withdraw from the cell cycle. *Exp. Cell Res.* **1987**, *169*, 215–222.

152. Schofield, P. N.; Engstrom, W.; Lee, A. J.; Biddle, C.; Graham, C. F. Expression of c-*myc* during differentiation of the human teratocarcinoma cell line Tera-2. *J. Cell Sci.* **1987**, *88*, 57–64.

153. Larsson, L-G.; Gray, H. E.; Totterman, T.; Pettersson, U.; Nilsson, K. Drastically increased expression of MYC and FOS protooncogenes during *in vitro* differentiation of chronic lymphocytic leukemia cells. *Proc. Natl. Acad. Sci. USA* **1987**, *84*, 223–227.

154. Freytag, S. O. Enforced expression of the c-*myc* oncogene inhibits cell differentiation by precluding entry into a distinct predifferentiation state in G0/G1. *Mol. Cell. Biol.* **1988**, *8*, 1614–1624.

155. Coppola, J. A.; Cole, M. D. Constitutive c-*myc* oncogene expression blocks mouse erythroleukaemia cell differentiation but not commitment. *Nature* **1986**, *320*, 760–763.

156. Lachman, H. M.; Cheng, G.; Skoultchi, A. I. Transfection of mouse erythroleukemia cells with *myc* sequences changes the rate of induced commitment to differentiate. *Proc. Natl. Acad. Sci. USA* **1986**, *83*, 6480–6484.

157. Dmitrovsky, E.; Kuehl, W. M.; Hollis, G. F.; Kirsch, I. R.; Bender, T. P.; Segal, S. Expression of a transfected human c-*myc* oncogene inhibits differentiation of a mouse erythroleukaemia cell line. *Nature* **1986**, *322*, 748–750.

158. Denis, N.; Blanc, S.; Leibovitch, M. P.; et al. c-*myc* oncogene expression inhibits the initiation of myogenic differentiation. *Exp. Cell. Res.* **1987**, *172*, 212–217.

159. Falcone, G.; Tato, F.; Alema, S. Distinctive effects of the viral oncogenes *myc*, *erb*, *fps*, and *src* on the differentiation program of quail myogenic cells. *Proc. Natl. Acad. Sci. USA* **1985**, *82*, 426–430.

160. Schneider, M. D.; Perryman, M. B.; Payne, P. A.; Spizz, G.; Roberts, R.; Olson, E. N. Autonomous expression of c-*myc* in BC3H1 cells partially inhibits but does not prevent myogenic differentiation. *Mol. Cell. Biol.* **1987**, *7*, 1973–1977.

161. Heikkila, R.; Schwab, G.; Wickstrom, E.; et al. A c-*myc* antisense oligodeoxynucleotide inhibits entry into S phase but not progress from G0 to G1. *Nature* **1987**, *328*, 445–449.

162. Griep, A. E.; Westphal, H. Antisense Myc sequences induce differentiation of F9 cells. *Proc. Natl. Acad. Sci. USA* **1988**, *85*, 6806–6810.

163. Prochownik, E. V.; Kukowska, J.; Rodgers, C. c-*myc* antisense transcripts accelerate differentiation and inhibit G1 progression in murine erythroleukemia cells. *Mol. Cell. Biol.* **1988**, *8*, 3683–3695.

164. Holt, J. T.; Redner, R. L.; Nienhuis, A. W. An oligomer complementary to c-*myc* mRNA inhibits proliferation of HL-60 promyelocytic cells and induces differentiation. *Mol. Cell. Biol.* **1988**, *8*, 963–973.

165. Bacon, T. A.; Wickstrom, E. Daily addition of an anti-c-*myc* DNA oligomer induces granulocytic differentiation of human promyelocytic leukemia HL-60 cells in both serum-containing and serum-free media. *Oncogene Res.* **1991**, *6*, 21–32.

166. Wickstrom, E. L.; Bacon, T. A.; Gonzalez, A.; Freeman, D. L.; Lyman, G. H.; Wickstrom, E. Human promyelocytic leukemia HL-60 cell proliferation and c-*myc* protein expression are

inhibited by an antisense pentadecadeoxynucleotide targeted against c-*myc* mRNA. *Proc. Natl. Acad. Sci. USA* **1988**, *85*, 1028–1032.

167. Hashiro, M.; Matsumoto, K.; Okumura, H.; Hashimoto, K.; Yoshikawa, K. Growth inhibition of human keratinocytes by antisense c-*myc* oligomer is not coupled to induction of differentiation. *Biochem. Biophys. Res. Commun.* **1991**, *174*, 287–292.

168. Watson, P. H.; Pon, R. T.; Shiu, R. P. C. Inhibition of c-*myc* expression by phosphorothioate antisense oligonucleotide identifies a critical role for c-*myc* in the growth of human breast cancer. *Cancer Res.* **1991**, *51*, 3996–4000.

169. Collins, J. F.; Herman, P.; Schuch, C.; Bagby, G. C. Jr. c-*myc* antisense oligonucleotides inhibit the colony-forming capacity of Colo 320 colonic carcinoma cells. *J. Clin. Invest.* **1992**, *89*, 1523–1527.

170. Biro, S.; Fu, Y-M.; Yu, Z-X.; Epstein, S. E. Inhibitory effects of antisense oligodeoxynucleotides targeting c-*myc* mRNA on smooth muscle cell proliferation and migration. *Proc. Natl. Acad. Sci. USA* **1993**, *90*, 654–658.

171. Sklar, M. D.; Thompson, E.; Welsh, M. J.; et al. Depletion of c-*myc* with specific antisense sequences reverses the transformed phenotype in *ras* oncogene-transformed NIH3T3 cells. *Mol. Cell. Biol.* **1991**, *11*, 3699–3710.

172. Wyllie, A. H. Glucocorticoid-induced thymocyte apoptosis is associated with endogenous nuclease activation. *Nature* **1980**, *284*, 555–556.

173. Askew, D. S.; Ashmun, R. A.; Simmons, B. C.; Cleveland, J. L. Constitutive c-*myc* expression in an IL3-dependent myeloid cell line suppresses cell cycle arrest and accelerates apotosis. *Oncogene* **1991**, *6*, 1915–1922.

174. Evan, G. I.; Wyllie, A. H.; Gilbert, C. S. et al. Induction of apoptosis in fibroblasts by c-*myc* protein. *Cell* **1992**, *69*, 119–128.

175. Bissonnette, R. P.; Echeverri, F.; Mahboubi, A.; Green, D. R. Apoptotic cell death induced by c-*myc* is inhibited by *bcl*-2. *Nature* **1992**, *359*, 552–554.

176. Fanidi, A.; Harrington, E. A.; Evan, G. I. Cooperative interaction between c-*myc* and *bcl*-2 protoonocogenes. *Nature* **1992**, *359*, 554–556.

177. Ames, B. N.; Gold, L. S. Chemical carcinogenesis: Too many rodent carcinogens. *Proc. Natl. Acad. Sci. USA* **1990**, *87*, 7772–7776.

178. Shi, Y.; Glynn, J. M.; Guilbert, L. J.; Cotter, T. G.; Bissonnette, R. P.; Green, D. R. Role of c-*myc* in activation-induced apoptotic cell death in T cell hybridomas. *Science* **1992**, *257*, 212–214.

179. Wurm, F. M.; Gwinn, K. A.; Kingston, R. E. Inducible overproduction of the mouse c-*myc* protein in mammalian cells. *Proc. Natl. Acad. Sci. USA* **1986**, *83*, 5414–5418.

180. Pallavicini, M. G.; Rosette, C.; Reitsma, M.; Deteresa, P. S.; Gray, J. W. Relationship of c-*myc* gene copy number and gene expression: Cellular effects of elevated c-*myc* protein. *J. Cell. Physiol.* **1990**, *143*, 372–380.

181. Wyllie, A. H.; Rose, K. A.; Morris, R. G.; Steel, C. M.; Foster, E.; Spandidos, D. A. Rodent fibroblast tumours expressing human *myc* and *ras* genes: Growth, metastasis and endogenous oncogene expression. *Biochem. Biophys. Res. Commun.* **1987**, *56*, 251–259.

182. Neiman, P. E.; Thomas, S. J.; Loring, G. Induction of apoptosis during normal and neoplastic B-cell development in the bursa of Fabricius. *Proc. Natl. Acad. Sci. USA* **1991**, *88*, 5857–5861.

183. Sandgren, E. P.; Quaife, C. J.; Paulovich, A. G.; Palmiter, R. D.; Brinster, R. L. Pancreatic tumor pathogenesis reflects the causative genetic lesion. *Proc. Natl. Acad. Sci. USA* **1991**, *88*, 93–97.

184. Quarmby, V. E.; Beckman, W. C. Jr.; Wilson, E. M.; French, F. S. Androgen regulation of c-*myc* messenger ribonucleic acid levels in rat ventral prostate. *Mol. Endocrinol.* **1987**, *1*, 865–874.

185. Buttyan, R.; Zakeri, Z.; Lockshin, R.; Wolgemuth, D. Cascade induction of c-*fos*, c-*myc*, and heat shock 70K transcripts during regression of the rat ventral prostate gland. *Mol. Endocrinol.* **1988**, *2*, 650–657.

186. Kyprianou, N.; English, H. F.; Davidson, N. E.; Isaacs, J. T. Programmed cell death during regression of the MCF-7 human breast cancer following estrogen ablation. *Cancer Res.* **1991**, *51*, 162–166.
187. Kelly, J. M.; Gilbert, C. S.; Stark, G. R.; Kerr, I. M. Differential regulation of interferon-induced mRNAs and c-*myc* mRNA by α- and γ-interferons. *Eur. J. Biochem.* **1985**, *153*, 367–371.
188. Wu, FY-H.; Chang, N-T.; Chen, W-J.; Juan, C-C. Vitamin K₃-induced cell cycle arrest and apoptotic cell death are accompanied by altered expression of c-*fos* and c-*myc* in nasopharyngeal carcinoma cells. *Oncogene* **1993**, *8*, 2237–2244.
189. Blackwell, T. K.; Kretzner, L.; Blackwood, E. M.; Eisenman, R. N.; Weintraub, H. Sequence-specific DNA binding by the c-*myc* protein. *Science* **1990**, *250*, 1149–1151.
190. Kerkhoff, E.; Bister, K.; Klempnauer, K-H. Sequence-specific DNA-binding by Myc proteins. *Proc. Natl. Acad. Sci. USA* **1991**, *88*, 4323–4327.
191. Prendergast, G. C.; Ziff, E. B. Methylation-sensitive sequence-specific DNA binding by the c-Myc basic region. *Science* **1991**, *251*, 186–189.
192. Kato, G. J.; Barrett, J.; Villa-Garcia, M.; Dang, C. V. An amino-terminal c-*myc* domain required for neoplastic transformation activates transcription. *Mol. Cell. Biol.* **1990**, *10*, 5914–5920.
193. Kretzner, L.; Blackwood, E. M.; Eisenman, R. N. Myc and Max proteins possess distinct transcriptional activities. *Nature* **1992**, *359*, 426–429.
194. Amati, B.; Dalton, S.; Brooks, M. W.; Littlewood, T. D.; Evan, G. I.; Land, H. Transcriptional activation by the human c-Myc oncoprotein in yeast requires interaction with Max. *Nature* **1992**, *359*, 423–426.
195. Blackwood, E. M.; Luscher, B.; Eisenman, R. N. Myc and Max associate *in vivo*. *Genes Dev.* **1992**, *6*, 71–80.
196. Blackwood, E. M.; Eisenman, R. N. Max: A helix-loop-helix zipper protein that forms a sequence-specific DNA-binding complex with *myc*. *Science* **1991**, *251*, 1211–1217.
197. Prendergast, G. C.; Lawe, D.; Ziff, E. B. Association of Myn, the murine homolog of Max, with c-*myc* stimulates methylation-sensitive DNA binding and *ras* cotransformation. *Cell* **1991**, *65*, 395–407.
198. Kato, G. J.; Lee, W. M. F.; Chen, L.; Dang, C. V. Max: functional domains and interaction with c-Myc. *Genes Dev.* **1992**, *6*, 81–92.
199. Gu, W.; Cechova, K.; Tassi, V.; Dalla-Favera, R. Opposite regulation of gene transcription and cell proliferation by c-Myc and Max. *Proc. Natl. Acad. Sci. USA* **1993**, *90*, 2935–2939.
200. Prendergast, G. C.; Hopewell, R.; Gorham, B. J.; Ziff, E. B. Biphasic effect of Max on Myc cotransformation activity and dependence on amino- and carboxy-terminal Max functions. *Genes Dev.* **1992**, *6*, 2429–2439.
201. Amin, C.; Wagner, A. J.; Hay, N. Sequence-specific transcriptional activation by *myc* and repression by max. *Mol. Cell. Biol.* **1993**, *13*, 383–390.
202. Gupta, S.; Seth, A.; Davis, R. J. Transactivation of gene expression by Myc is inhibited by mutation at the phosphorylation sites Thr-58 and Ser-62. *Proc. Natl. Acad. Sci. USA* **1993**, *90*, 3216–3220.
203. Eilers, M.; Schirm, S.; Bishop, J. M. The *myc* protein activates transcription of the α-prothymosin gene. *EMBO J.* **1991**, *10*, 133–141.
204. Bello-Fernandez, C.; Cleveland, J. L. c-*myc* transactivates the ornithine decarboxylase gene. *Curr. Topic Microbiol. & Immunol.* **1992**, *182*, 445–452.
205. Reisman, D.; Elkind, N. B.; Roy, B.; Beamon, J.; Rotter, V. c-*myc* trans-activates the p53 promoter through a required downstream CACGTG motif. *Cell Growth & Differ.* **1993**, *4*, 57–65.
206. Kaddurah-Daouk, R.; Greene, J. M.; Baldwin, A. S. Jr.; Kingston, R. E. Activation and repression of mammalian gene expression by the c-*myc* protein. *Genes Dev.* **1987**, *1*, 347–357.
207. Prendergast, G. C.; Diamond, L. E.; Dahl, D.; Cole, M. D. The c-*myc*-regulated gene *mr1* encodes plasminogen activator inhibitor 1. *Mol. Cell. Biol.* **1990**, *10*, 1265–1269.

208. Jansen-Durr, P.; Meichle, A.; Steiner, P.; et al. Differential modulation of cyclin gene expression by MYC. *Proc. Natl. Acad. Sci. USA* **1993**, *90*, 3685–3689.
209. Versteeg, R.; Noordermeer, I. A.; Kruse-Wolters, M.; Ruiter, D. J.; Schrier, P. I. c-*myc* down-regulates class I HLA expression in human melanomas. *EMBO J.* **1988**, *7*, 1023–1029.
210. Inghirami, G.; Grignani, F.; Sternas, L.; Lombardi, L.; Knowles, D. M.; Dalla-Favera, R. Downregulation of LFA-1 adhesion receptors by c-*myc* in human B lymphoblastoid cells. *Science* **1990**, *250*, 682–686.
211. Cheng, G.; Skoultchi, A. I. Rapid induction of polyadenylated H1 histone mRNAs in mouse erythroleukemia cells is regulated by c-*myc*. *Mol. Cell. Biol.* **1989**, *9*, 2332–2340.
212. Yang, B-S.; Geddes, T. J.; Pogulis, R. J.; de Crombrugghe, B.; Freytag, S. O. Transcriptional suppression of cellular gene expression by c-*myc*. *Mol. Cell. Biol.* **1991**, *11*, 2291–2295.
213. Yang, B-S.; Gilbert, J. D.; Freytag, S. O. Overexpression of *myc* suppresses CCAAT transcription factor/nuclear factor 1-dependent promoters *in vivo*. *Mol. Cell. Biol.* **1993**, *13*, 3093–3102.
214. Prendergast, G. C.; Cole, M. D. Posttranscriptional regulation of cellular gene expression by the c-*myc* oncogene. *Mol. Cell. Biol.* **1989**, *9*, 124–134.
215. Gibson, A. W.; Ye, R.; Johnston, R. N.; Browder, L. W. A possible role for c-*myc* oncoproteins in post-transcriptional regulation of ribosomal RNA. *Oncogene* **1992**, *7*, 2363–2367.
216. Godeau, F.; Persson, H.; Gray, H. E.; Pardee, A. B. c-*myc* expression is dissociated from DNA synthesis and cell division in Xenopus oocyte and early embryonic development. *EMBO J.* **1986**, *5*, 3571–3577.
217. Penn, L. J. Z.; Brooks, M. W.; Laufer, E. M.; Land, H. Negative autoregulation of c-*myc* transcription. *EMBO J.* **1990**, *9*, 1113–1121.
218. Murray, J. M.; Nishikura, K. The mechanism of inactivation of the normal c-*myc* gene locus in human Burkitt lymphoma cells. *Oncogene* **1987**, *2*, 493–498.
219. Mango, S. E.; Schuler, G. D.; Steele, M. E. R.; Cole, M. D. Germ line c-*myc* is not down-regulated by loss or exclusion of activating factors in *myc*-induced macrophage tumors. *Mol. Cell. Biol.* **1989**, *9*, 3482–3490.
220. Alexander, W. S.; Schrader, J. W.; Adams, J. M. Expression of the c-*myc* oncogene under control of an immunoglobulin enhancer in Eμ-*myc* transgenic mice. *Mol. Cell. Biol.* **1987**, *7*, 1436–1444.
221. Cleveland, J. L.; Huleihel, M.; Bressler, P.; et al. Negative regulation of c-*myc* transcription involves *myc* family proteins. *Oncogene Res.* **1988**, *3*, 357–375.
222. Kitaura, H.; Galli, I.; Taira, T.; Iguchi-Ariga, S. M. M.; Ariga, H. Activation of c-*myc* promoter by c-*myc* protein in serum starved cells. *FEBS Lett.* **1991**, *290*, 147–152.
223. Iguchi-Ariga, S. M. M.; Okazaki, T.; Itani, T.; Ogata, M.; Sato, Y.; Ariga, H. An initiation site of DNA replication with transcriptional enhancer activity present upstream of the c-*myc* gene. *EMBO J.* **1988**, *7*, 3135–3143.
224. Stone, J.; de Lange, T.; Ramsay, G.; et al. Definition of regions in human c-*myc* that are involved in transformation and nuclear localization. *Mol. Cell. Biol.* **1987**, *7*, 1697–1709.
225. Sarid, J.; Halazonetis, T. D.; Murphy, W.; Leder, P. Evolutionarily conserved regions of the human c-*myc* protein can be uncoupled from transforming activity. *Proc. Natl. Acad. Sci. USA* **1987**, *84*, 170–173.
226. Penn, L. J. Z.; Brooks, M. W.; Laufer, E. M.; et al. Domains of human c-*myc* protein required for autosuppression and cooperation with *ras* oncogenes are overlapping. *Mol. Cell. Biol.* **1990**, *10*, 4961–4966.
227. Freytag, S. O.; Dang, C. V.; Lee, W. M. F. Definition of the activities and properties of c-*myc* required to inhibit cell differentiation. *Cell Growth & Differ.* **1990**, *1*, 339–343.
228. McKenna, W. G.; Weiss, M. C.; Endlich, B.; et al. Synergistic effect of the v-*myc* oncogene with H-*ras* on radioresistance. *Cancer Res.* **1990**, *50*, 97–102.
229. Chang, E. H.; Pirollo, K. F.; Zou, Z. Q.; et al. Oncogenes in radioresistant, noncancerous skin fibroblasts from a cancer-prone family. *Science* **1987**, *237*, 1036–1039.

230. Little, C. D.; Nau, M. M.; Carney, D. N.; Gazdar, A. F.; Minna, J. D. Amplification and expression of the c-*myc* oncogene in human lung cancer cell lines. *Nature* **1983**, *306*, 194–196.

231. Carney, D. N.; Mitchell, J. B.; Kinsella, T. J. *In vitro* radiation and chemotherapy sensitivity of established cell lines of human small cell lung cancer and its large cell morphological variants. *Cancer Res.* **1983**, *43*, 2806–2811.

232. Denis, N.; Kitzis, A.; Kruh, J.; Dautry, F.; Corcos, D. Stimulation of methotrexate resistance and dihydrofolate reductase gene amplification by c-*myc*. *Oncogene* **1991**, *6*, 1453–1457.

233. Cerni, C.; Mougneau, E.; Cuzin, F. Transfer of 'immortalizing' oncogenes into rat fibroblasts induces both high rates of sister chromatid exchange and appearance of abnormal karyotypes. *Exp. Cell. Res.* **1987**, *168*, 439–446.

234. Niimi, S.; Nakagawa, K.; Yokota, J.; et al. Resistance to anticancer drugs in NIH3T3 cells transfected with c-*myc* and/or c-H-*ras* genes. *Br. J. Cancer* **1991**, *63*, 237–241.

235. Sklar, M. D.; Prochownik, E. V. Modulation of cis-platinum resistance in Friend erythroleukemia cells by c-*myc*. *Cancer Res.* **1991**, *51*, 2118–2123.

236. Delaporte, C.; Larsen, A. K.; Dautry, F.; Jacquemin-Sablon, A. Influence of *myc* overexpression on the phenotypic properties of Chinese hamster lung cells resistant to antitumor agents. *Exp. Cell Res.* **1991**, *197*, 176–182.

237. Gazdar, A. F.; Carney, D. N.; Nau, M. M.; Minna, J. D. Characterization of variant subclasses of cell lines derived from small cell lung cancer having distinctive biochemical, morphological, and growth properties. *Cancer Res.* **1985**, *45*, 2924–2930.

238. Saksela, K.; Bergh, J.; Lehto, V-P.; Nilsson, K.; Alitalo, K. Amplification of the c-*myc* oncogene in a subpopulation of human small cell lung cancer. *Cancer Res.* **1985**, *45*, 1823–1827.

239. Kiefer, P. E.; Bepler, G.; Kubasch, M.; Havemann, K. Amplification and expression of protoon-cogenes in human small cell lung cancer cell lines. *Cancer Res.* **1987**, *47*, 6236–6242.

240. Johnson, B. E.; Ihde, D. C.; Makuch, R. W.; et al. *myc* family oncogene amplification in tumor cell lines established from small cell lung cancer patients and its relationship to clinical status and course. *J. Clin. Invest.* **1987**, *79*, 1629–1634.

241. Brennan, J.; O'Connor, T.; Makuch, R.W.; et al. *myc* family DNA amplification in 107 tumors and tumor cell lines from patients with small cell lung cancer treated with different combination chemotherapy regimens. *Cancer Res.* **1991**, *51*, 1708–1712.

242. Yokota, J.; Wada, M.; Yoshida, T.; et al. Heterogeneity of lung cancer cells with respect to the amplification and rearrangement of *myc* family oncogenes. *Oncogene* **1988**, *2*, 607–611.

243. Takahashi, T.; Obata, Y.; Sekido, Y.; et al. Expression and amplification of *myc* gene family in small cell lung cancer and its relation to biological characteristics. *Cancer Res.* **1989**, *49*, 2683–2688.

244. Johnson, B. E.; Makuch, R. W.; Simmons, A. D.; Gazdar, A. F.; Burch, D.; Cashell, A. W. *myc* family DNA amplification in small cell lung cancer patients' tumors and corresponding cell lines. *Cancer Res.* **1988**, *48*, 5163–5166.

245. Wong, A. J.; Ruppert, J. M.; Eggleston, J.; Hamilton, S. R.; Baylin, S. B.; Vogelstein, B. Gene amplification of c-*myc* and N-*myc* in small cell carcinoma of the lung. *Science* **1986**, *233*, 461–464.

246. Gosney, J. R.; Field, J. K.; Gosney, M. A.; Lye, M. D. W.; Spandidos, D. A.; Butt, S. A. c-*myc* oncoprotein in bronchial carcinoma: expression in all major morphological types. *Anticancer Res.* **1990**, *10*, 623–628.

247. Gu, J.; Linnoila, R. I.; Seibel, N. L.; et al. A study of *myc*-related gene expression in small cell lung cancer by in situ hybridization. *Am. J. Pathol.* **1988**, *132*, 13–17.

248. Krystal, G.; Birrer, M.; Way, J.; et al. Multiple mechanisms for transcriptional regulation of the *myc* gene family in small-cell lung cancer. *Mol. Cell Biol.* **1988**, *8*, 3373–3381.

249. Rygaard, K.; Vindelov, L. L.; Spang-Thomsen, M. Expression of *myc* family oncoproteins in small-cell lung-cancer cell lines and xenografts. *Int. J. Cancer* **1993**, *54*, 144–152.

250. Nau, M. M.; Brooks, B. J. Jr.; Carney, D. N.; et al. Human small-cell lung cancer cell lines show amplification and expression of the N-*myc* gene. *Proc. Natl. Acad. Sci. USA* **1986**, *83*, 1092–1096.

251. Cline, M. J.; Battifora, H. Abnormalities of protooncogenes in non-small cell lung cancer. *Cancer* **1987**, *60*, 2669–2674.

252. Gemma, A.; Nakajima, T.; Shiraishi, M.; et al. *myc* family gene abnormality in lung cancers and its relation to xenotransplantability. *Cancer Res.* **1988**, *48*, 6025–6028.

253. Spandidos, D. A.; Zakinthinos, S.; Petraki, C.; et al. Expression of *ras* p21 and *myc* p62 oncoproteins in small cell and non small cell carcinoma of the lung. *Anticancer Res.* **1990**, *10*, 1105–1114.

254. Broers, J. L. V.; Viallet, J.; Jensen, S. M.; et al. Expression of c-*myc* in progenitor cells of the bronchopulmonary epithelium and in a large number of non-small cell lung cancers. *Am. J. Respir. Cell Mol. Biol.* **1993**, *9*, 33–43.

255. Yoshimoto, K.; Hirohashi, S.; Sekiya, T. Increased expression of the c-*myc* gene without gene amplification in human lung cancer and colon cancer cell lines. *Jpn. J. Cancer Res.* **1986**, *77*, 540–545.

256. Yoshimoto, K.; Shiraishi, M.; Hirohashi, S.; et al. Rearrangement of the c-*myc* gene in two giant cell carcinomas of the lung. *Jpn. J. Cancer Res.* **1986**, *77*, 731–735.

257. Iizuka, M.; Shiraishi, M.; Yoshida, M. C.; Hayashi, K.; Sekiya, T. Joining of the c-*myc* gene and a Line 1 family member on chromosome 8 in a human primary giant cell carcinoma of the lung. *Cancer Res.* **1990**, *50*, 3345–3350.

258. Sundaresan, V.; Reeve, J. G.; Wilson, B.; Bleehen, N. M.; Watson, J. V. Flow cytometric and immunohistochemical analysis of p62$^{c\text{-}myc}$ oncoprotein in the bronchial epithelium of lung cancer patients. *Anticancer Res.* **1991**, *11*, 2111–2116.

259. Johnson, B. E.; Battey, J.; Linnoila, I.; et al. Changes in the phenotype of human small cell lung cancer cell lines after transfection and expression of the c-*myc* proto-oncogene. *J. Clin. Invest.* **1986**, *78*, 525–532.

260. Pfeifer, A. M. A.; Mark, G. E. III, Malan-Shibley, L.; Graziano, S.; Amstad, P.; Harris, C. C. Cooperation of c-*raf*-1 and c-*myc* protooncogenes in the neoplastic transformation of simian virus 40 large tumor antigen-immortalized human bronchial epithelial cells. *Proc. Natl. Acad. Sci. USA* **1989**, *86*, 10075–10079.

261. Pfeifer, A. M. A.; Jones, R. T.; Bowden, P. E.; et al. Human bronchial epithelial cells transformed by the c-*raf*-1 and c-*myc* protooncogenes induce multidifferentiated carcinomas in nude mice: A model for lung carcinogenesis. *Cancer Res.* **1991**, *51*, 3793–3801.

262. Clement, A.; Campisi, J.; Farmer, S. R.; Brody, J. S. Constitutive expression of growth-related mRNAs in proliferating and nonproliferating lung epithelial cells in primary culture: Evidence for growth-dependent translational control. *Proc. Natl. Acad. Sci. USA* **1990**, *87*, 318–322.

263. Bepler, G.; Carney, D. N.; Nau, M. M.; Gazdar, A. F.; Minna, J. D. Additive and differential biological activity of α-interferon A, difluoromethylornithine, and their combination on established human lung cancer cell lines. *Cancer Res.* **1986**, *46*, 3413–3419.

264. Dani, C.; Mechti, M.; Piechaczyk, M.; Lebleu, B.; Jeanteur, P.; Blanchard, J. M. Increased rate of degradation of c-*myc* mRNA in interferon-treated Daudi cells. *Proc. Natl. Acad. Sci. USA* **1985**, *82*, 4896–4899.

265. Einat, M.; Resnitzky, D.; Kimchi, A. Close link between reduction of c-*myc* expression by interferon and G0/G1 arrest. *Nature* **1985**, *313*, 597–600.

266. Kozbor, D.; Croce, C. M. Amplification of the c-*myc* oncogene in one of five human breast carcinoma cell lines. *Cancer Res.* **1984**, *44*, 438–441.

267. Modjtahedi, N.; Lavialle, C.; Poupon, M-F.; et al. Increased level of amplification of the c-*myc* oncogene in tumors induced in nude mice by a human breast carcinoma cell line. *Cancer Res.* **1985**, *45*, 4372–4379.

268. Dubik, D.; Dembinski, T. C.; Shiu, R. P. C. Stimulation of c-*myc* oncogene expression associated with estrogen-induced proliferation of human breast cancer cells. *Cancer Res.* **1987**, *47*, 6517–6521.

269. Krief, P.; Saint-Ruf, C.; Bracke, M.; et al. Acquisition of tumorigenic potential in the human myoepithelial HBL100 cell line is associated with decreased expression of HLA class I, class II and integrin β3 and increased expression of c-*myc*. *Int. J. Cancer* **1989**, *43*, 658–664.

270. Collyn-d'Hooghe, M.; Vandewalle, B.; Hornez, L.; et al. C-*myc* overexpression, c-*mil*, c-*myb* expression in a breast tumor cell line. Effects of estrogen and antiestrogen. *Anticancer Res.* **1991**, *11*, 2175–2180.

271. Watanabe, M.; Tanaka, H.; Kamada, M.; et al. Establishment of the human BSMZ breast cancer cell line, which overexpresses the *erb*B-2 and c-*myc* genes. *Cancer Res.* **1992**, *52*, 5178–5182.

272. Berns. E. M. J. J.; Foekens, J. A.; van Putten, W. L. J.; et al. Prognostic factors in human primary breast cancer: comparison of c-*myc* and HER2/*neu* amplification. *J. Steroid Biochem. Molec. Biol.* **1992**, *43*, 13–19.

273. Roux-Dosseto, M.; Romain, S.; Dussault, N.; et al. c-*myc* gene amplification in selected node-negative breast cancer patients correlates with high rate of early relapse. *Eur. J. Cancer* **1992**, *28A*, 1600–1604.

274. Seshadri, R.; Matthews, C.; Dobrovic, A.; Horsfall, D. J. The significance of oncogene amplification in primary breast cancer. *Int. J. Cancer* **1989**, *43*, 270–272.

275. Tang, R.; Kacinski, B.; Validire, P.; et al. Oncogene amplification correlates with dense lymphocyte infiltration in human breast cancers: A role for hematopoietic growth factor release by tumor cells? *J. Cell Biochem.* **1990**, *44*, 189–198.

276. Brouillet, J-P.; Theillet, C.; Maudelonde, T.; et al. Cathepsin D assay in primary breast cancer and lymph nodes: Relationship with c-*myc*, c-*erb*-B-2 and int-2 oncogene amplification and node invasiveness. *Eur. J. Cancer* **1990**, *26*, 437–441.

277. Varley, J. M.; Swallow, J. E.; Brammar, W. J.; Whittaker, J. L.; Walker, R. A. Alterations to either c-*erb*B-2 (*neu*) or c-*myc* proto-oncogenes in breast carcinomas correlate with poor short-term prognosis. *Oncogene* **1987**, *1*, 423–430.

278. Mariani-Constantini, R.; Escot, C.; Theillet, C.; et al. In situ c-*myc* expression and genomic status of the c-*myc* locus in infiltrating ductal carcinomas of the breast. *Cancer Res.* **1988**, *48*, 199–205.

279. Escot, C.; Theillet, C.; Lidereau, R.; et al. Genetic alteration of the c-*myc* protooncogene (MYC) in human breast carcinomas. *Proc. Natl. Acad. Sci. USA* **1986**, *83*, 4834–4838.

280. Biunno, I.; Pozzi, M. R.; Pierotti, M. A.; Pilotti, S.; Cattoretti, G.; Della Porta, G. Structure and expression of oncogenes in surgical specimens of human breast carcinomas. *Br. J. Cancer* **1988**, *57*, 464–468.

281. Bonilla, M.; Ramirez, M.; Lopez-Cueto, J.; Gariglio, P. In vivo amplification and rearrangement of c-*myc* oncogene in human breast tumors. *J. Natl. Cancer Inst.* **1988**, *80*, 665–671.

282. Berns, E. M. J. J.; Klijn, J. G. M.; van Staveren, I. L.; Portengen, H.; Noordegraaf, E.; Foekens, J. A. Prevalence of amplification of the oncogenes c-*myc*, HER2/*neu*, and *int*-2 in one thousand human breast tumours: Correlation with steroid receptors. *Eur. J. Cancer* **1992**, *28*, 697–700.

283. Morse, B.; Rothberg, P. G.; South, V. J.; Spandorfer, J. M.; Astrin, S. M. Insertional mutagenesis of the *myc* locus by a LINE-1 sequence in a human breast carcinoma. *Nature* **1988**, *333*, 87–90.

284. Borg, Å.; Baldetorp, B.; Ferno, M.; Olsson, H.; Sigurdsson, H. c-*myc* amplification is an independent prognostic factor in postmenopausal breast cancer. *Int. J. Cancer* **1992**, *51*, 687–691.

285. Berns, E. M. J. J.; Klijn, J. G. M.; van Putten, W. L. J.; van Staveren, I. L.; Portengen, H.; Foekens, J. A. c-*myc* amplification is a better prognostic factor than *HER2/neu* amplification in primary breast cancer. *Cancer Res.* **1992**, *52*, 1107–1113.

286. Gutman, M.; Ravia, Y.; Assaf, D.; Yamamoto, T.; Rozin, R.; Shiloh, Y. Amplification of c-*myc* and c-*erb*B-2 proto-oncogenes in human solid tumors: frequency and clinical significance. *Int. J. Cancer* **1989**, *44*, 802–805.

287. Cline, M. J.; Battifora, H.; Yokota, J. Proto-oncogene abnormalities in human breast cancer: correlations with anatomic features and clinical course of disease. *J. Clin. Oncol.* **1987**, *5*, 999–1006.
288. Garcia, I.; Dietrich, P-Y.; Aapro, M.; Vauthier, G.; Vadas, L.; Engel, E. Genetic alterations of c-*myc*, c-*erb*B-2, and c-Ha-*ras* protooncogenes and clinical associations in human breast carcinomas. *Cancer Res.* **1989**, *49*, 6675–6679.
289. Meyers, S. L.; O'Brien, M. T.; Smith, T.; Dudley, J. P. Analysis of the *int*-1, *int*-2, c-*myc*, and *neu* oncogenes in human breast carcinomas. *Cancer Res.* **1990**, *50*, 5911–5918.
290. Guérin, M.; Barrois, M.; Terrier, M-J.; Spielmann, M.; Riou, G. Overexpression of either c-*myc* or c-*erb*-2/*neu* proto-oncogenes in human breast carcinomas: correlation with poor prognosis. *Oncogene Res.* **1988**, *3*, 21–31.
291. Adnane, J.; Gaudray, P.; Simon, M-P.; Simony-Lafontaine, J.; Jeanteur, P.; Theillet, C. Protooncogene amplification and human breast tumor phenotype. *Oncogene* **1989**, *4*, 1389–1395.
292. Donovan-Peluso, M.; Contento, A. M.; Tobon, H.; Ripepi, B.; Locker, J. Oncogene amplification in breast cancer. *Am. J. Pathol.* **1991**, *138*, 835–845.
293. Ottestad, L.; Andersen, T. I.; Nesland, J. M. et al. Amplification of c-*erb*B-2, int-2, and c-*myc* genes in node-negative breast carcinomas. *Acta Oncologica.* **1993**, *32*, 289–294.
294. Pavelic, Z. P.; Steele, P.; Preisler, H. D. Evaluation of c-*myc* proto-oncogene in primary human breast carcinomas. *Anticancer Res.* **1991**, *11*, 1421–1428.
295. Pavelic, Z. P.; Pavelic, K.; Carter, C. P.; Pavelic, L. Heterogeneity of c-*myc* expression in histologically similar infiltrating ductal carcinomas of the breast. *J. Cancer Res. Clin. Oncol.* **1992**, *118*, 16–22.
296. Pavelic, Z. P.; Pavelic, L.; Lower, E. E.; et al. c-*myc*, c-*erb*B-2, and Ki-67 expression in normal breast tissue and in invasive and noninvasive breast carcinoma. *Cancer Res.* **1992**;, *52*, 2597–2602.
297. Whittaker, J. L.; Walker, R. A.; Varley, J. M. Differential expression of cellular oncogenes in benign and malignant human breast tissue. *Int. J. Cancer* **1986**, *38*, 651–655.
298. Locker, A. P.; Dowle, C. S.; Ellis, I. O.; et al. C-*myc* oncogene product expression and prognosis in operable breast cancer. *Br. J. Cancer* **1989**, *60*, 669–672.
299. Spandidos, D. A.; Field, J. K.; Agnantis, N. J.; Evan, G. I.; Moore, J. P. High levels of c-*myc* protein in human breast tumours determined by a sensitive ELISA technique. *Anticancer Res.* **1989**, *9*, 821–826.
300. Zajchowski, D.; Band, V.; Pauzie, N.; Taga, A.; Stampfer, M.; Sager, R. Expression of growth factors and oncogenes in normal and tumor-derived human mammary epithelial cells. *Cancer Res.* **1988**, *48*, 7041–7047.
301. Varley, J. M.; Wainwright, A. M.; Brammar, W. J. An unusual alteration in c-*myc* in tissue from a primary breast carcinoma. *Oncogene* **1987**, *1*, 431–438.
302. Katzir, N.; Rechavi, G.; Cohen, J. B.; et al. "Retroposon" insertion into the cellular oncogene c-*myc* in canine transmissible venereal tumor. *Proc. Natl. Acad. Sci. USA* **1985**, *82*, 1054–1058.
303. Kreipe, H.; Feist, H.; Fischer, L.; et al. Amplification of c-*myc* but not of c-*erb*B-2 is associated with high proliferative capacity in breast cancer. *Cancer Res.* **1993**, *53*, 1956–1961.
304. Dowle, C. S.; Robins, R. A.; Watkins, K.; Blamey, R. W.; Sikora, K.; Evans, G. I. The relationship of p62^{c-myc} in operable breast cancer to patient survival and tumour prognostic factors. *Br. J. Surg.* **1987**, *74*, 534.
305. Pertschuk, L. P.; Feldman, J. G.; Kim, D. S.; et al. Steroid hormone receptor immunohistochemistry and amplification of c-*myc* protooncogene. *Cancer* **1993**, *71*, 162–171.
306. Papamichalis, G.; Francia, K.; Karachaliou, F. E.; Anastasiades, O. T.; Spandidos, D. A. Expression of the c-*myc* oncoprotein in human metaplastic epithelial cells of fibrocystic disease. *Anticancer Res.* **1988**, *8*, 1217–1222.

307. Escot, C.; Simony-Lafontaine, J.; Maudelonde, T.; Puech, C.; Pujol, H.; Rochefort, H. Potential value of increased MYC but not *ERBB*2 RNA levels as a marker of high-risk mastopathies. *Oncogene* **1993**, *8*, 969–974.

308. Lavialle, C.; Modjtahedi, N.; Lamonerie, T.; et al. The human breast carcinoma cell line SW 613-S: an experimental system to study tumor heterogeneity in relation to c-*myc* amplification, growth factor production and other markers (review). *Anticancer Res.* **1989**, *9*, 1265–1280.

309. Lavialle, C.; Modjtahedi, N.; Cassingena, R.; Brison, O. High c-*myc* amplification level contributes to the tumorigenic phenotype of the human breast carcinoma cell line SW 613-S. *Oncogene* **1988**, *3*, 335–339.

310. Telang, N. T.; Osborne, M. P.; Sweterlitsch, L. A.; Narayanan, R. Neoplastic transformation of mouse mammary epithelial cells by deregulated *myc* expression. *Cell Regul.* **1990**, *1*, 863–872.

311. Ball, R. K.; Ziemiecki, A.; Schönenberger, C. A.; Reichmann, E.; Redmond, S. M. S.; Groner, B. v-*myc* alters the response of a cloned mouse epithelial cell line to lactogenic hormones. *Mol. Endocrinol.* **1988**, *2*, 133–142.

312. Stewart, T. A.; Pattengale, P. K.; Leder, P. Spontaneous mammary adenocarcinomas in transgenic mice that carry and express MTV/*myc* fusion genes. *Cell* **1984**, *38*, 627–637.

313. Schoenenberger, C-A.; Andres, A-C.; Groner, B.; van der Valk, M.; LeMeur, M.; Gerlinger, P. Targeted c-*myc* gene expression in mammary glands of transgenic mice induces mammary tumours with constitutive milk protein gene transcription. *EMBO J.* **1988**, *7*, 169–175.

314. Leder, A.; Pattengale, P. K.; Kuo, A.; Stewart, T. A.; Leder, P. Consequences of widespread deregulation of the c-*myc* gene in transgenic mice: Multiple neoplasms and normal development. *Cell* **1986**, *45*, 485–495.

315. Sinn, E.; Muller, W.; Pattengale, P.; Tepler, T.; Wallace, R.; Leder, P. Coexpression of MMTV/v-Ha-*ras* and MMTV/c-*myc* genes in transgenic mice: Synergistic action of oncogenes in vivo. *Cell* **1987**, *49*, 465–475.

316. Andres, A-C.; van der Valk, M. A.; Schonenberger, C-A.; et al. Ha-*ras* and c-*myc* oncogene expression interferes with morphological and functional differentiation of mammary epithelial cells in single and double transgenic mice. *Genes Dev.* **1988**, *2*, 1486–1495.

317. Dotto, G. P.; Weinberg, R. A.; Ariza, A. Malignant transformation of mouse primary keratinocytes by harvey sarcoma virus and its modulation by surrounding normal cells. *Proc. Natl. Acad. Sci. USA* **1988**, *85*, 6389–6393.

318. Bignami, M.; Rosa, S.; La Rocca, S. A.; Falcone, G.; Tato, F. Differential influence of adjacent normal cells on the proliferation of mammalian cells transformed by the viral oncogenes *myc*, *ras* and *src*. *Oncogene* **1988**, *3*, 509–514.

319. Edwards, P. A. W.; Ward, J. L.; Bradbury, J. M. Alteration of morphogenesis by the v-*myc* oncogene in transplants of mammary gland. *Oncogene* **1988**, *3*, 407–412.

320. Bradbury, J. M.; Sykes, H.; Edwards, P. A. W. Induction of mouse mammary tumours in a transplantation system by the sequential introduction of the MYC and RAS oncogenes. *Int. J. Cancer* **1991**, *48*, 908–915.

321. Morse, H. C. III; Hartley, J. W.; Fredrickson, T. N.; et al. Recombinant murine retroviruses containing avian v-*myc* induce a wide spectrum of neoplasms in newborn mice. *Proc. Natl. Acad. Sci. USA* **1986**, *83*, 6868–6872.

322. Santos, G. F.; Scott, G. K.; Lee, W. M. F.; Liu, E.; Benz, C. Estrogen-induced post-transcriptional modulation of c-*myc* proto-oncogene expression in human breast cancer cells. *J. Biol. Chem.* **1988**, *263*, 9565–9568.

323. Dubik, D.; Shiu, R. P. C. Transcriptional regulation of c-*myc* oncogene expression by estrogen in hormone-responsive human breast cancer cells. *J. Biol. Chem.* **1988**, *263*, 12705–12708.

324. van der Burg, B.; van Selm-Miltenburg, A. J. P.; de Laat, S. W.; van Zoelen, E. J. J. Direct effects of estrogen on c-*fos* and c-*myc* protooncogene expression and cellular proliferation in human breast cancer cells. *Mol. Cell. Endocrinol.* **1989**, *64*, 223–228.

325. Liu, E.; Santos, G.; Lee, W. M. F.; Osborne, C. K.; Benz, C. C. Effects of c-myc overexpression on the growth characteristics of MCF-7 human breast cancer cells. *Oncogene* **1989**, *4*, 979–984.

326. Wong, M. S. J.; Murphy, L. C. Differential regulation of c-myc by progestins and antiestrogens in T-47D human breast cancer cells. *J. Steroid Biochem. Mol. Biol.* **1991**, *39*, 39–44.

327. Wosikowski, K.; Küng, W.; Hasmann, M.; Loser, R.; Eppenberger, U. Inhibition of growth factor-activated proliferation by anti-estrogens and effects on early gene expression of MCF-7 cells. *Int. J. Cancer* **1993**, *53*, 290–297.

328. Le Roy, X.; Escot, C.; Brouillet, J-P.; et al. Decrease of c-erbB-2 and c-myc RNA levels in tamoxifen-treated breast cancer. *Oncogene* **1991**, *6*, 431–437.

329. Musgrove, E. A.; Lee, C. S. L.; Sutherland, R. L. Progestins both stimulate and inhibit breast cancer cell cycle progression while increasing expression of transforming growth factor α, epidermal growth factor receptor, c-fos, and c-myc genes. *Mol. Cell. Biol.* **1991**, *11*, 5032–5043.

330. Hamburger, A. W.; Pinnamaneni, G. Interferon induced increases in c-myc expression in a human breast carcinoma cell line. *Anticancer Res.* **1991**, *11*, 1891–1894.

331. Fernandez-Pol, J. A.; Talkad, V. D.; Klos, D. J.; Hamilton, P. D. Suppression of the EGF-dependent induction of c-myc proto-oncogene expression by transforming growth factor β in a human breast carcinoma cell line. *Biochem. Biophys. Res. Commun.* **1987**, *144*, 1197–1205.

332. Ocadiz, R.; Sauceda, R.; Cruz, M.; Graef, A. M.; Gariglio, P. High correlation between molecular alterations of the c-myc oncogene and carcinoma of the uterine cervix. *Cancer Res.* **1987**, *47*, 4173–4177.

333. Riou, G. F.; Bourhis, J.; Le, M. G. The c-myc proto-oncogene in invasive carcinomas of the uterine cervix: Clinical relevance of overexpression in early stages of the cancer. *Anticancer Res.* **1990**, *10*, 1225–1232.

334. Choo, K-B.; Chong, K-Y.; Chou, H-F.; Liew, L-N.; Liou, C-C. Analysis of the structure and expression of the c-myc oncogene in cervical tumor and in cervical tumor-derived cell lines. *Biochem. Biophys. Res. Commun.* **1989**, *158*, 334–340.

335. Hendy-Ibbs, P.; Cox, H.; Evan, G. I.; Watson, J. V. Flow cytometric quantitation of DNA and c-myc oncoprotein in archival biopsies of uterine cervix neoplasia. *Br. J. Cancer* **1987**, *55*, 275–282.

336. Covington, M.; Sikora, K.; Turner, M. J.; White, J. O.; Moore, P.; Soutter, W. P. c-myc expression in cervical cancer. *Lancet* **1987**, *i*, 1260–1261.

337. Symonds, R. P.; Habeshaw, T.; Paul, J.; et al. No correlation between ras, c-myc and c-jun protooncogene expression and prognosis in advanced carcinoma of cervix. *Eur. J. Cancer* **1992**, *28A*, 1615–1617.

338. Riou, G.; Barrois, M.; Le, M. G.; George, M.; Le Doussal, V.; Haie, C. C-myc proto-oncogene expression and prognosis in early carcinoma of the uterine cervix. *Lancet* **1987**, *8536*, 761–763.

339. Bourhis, J.; Le, M. G.; Barrois, M.; et al. Prognostic value of c-myc proto-oncogene overexpression in early invasive carcinoma of the cervix. *J. Clin. Oncol.* **1990**, *8*, 1789–1796.

340. Borst, M. P.; Baker, V. V.; Dixon, D.; Hatch, K. D.; Shingleton, H. M.; Miller, D. M. Oncogene alterations in endometrial carcinoma. *Gynecol. Oncol.* **1990**, *38*, 364–366.

341. Sasano, H.; Comerford, J.; Wilkinson, D. S.; Schwartz, A.; Garrett, C. T. Serous papillary adenocarcinoma of the endometrium. *Cancer* **1990**, *65*, 1545–1551.

342. Baker, V. V.; Borst, M. P.; Dixon, D.; Hatch, K. D.; Shingleton, H. M.; Miller, D. c-myc amplification in ovarian cancer. *Gynecol. Oncol.* **1990**, *38*, 340–342.

343. Sasano, H.; Garrett, C. T.; Wilkinson, D. S.; Silverberg, S.; Comerford, J.; Hyde, J. Protooncogene amplification and tumor ploidy in human ovarian neoplasms. *Hum. Pathol.* **1990**, *21*, 382–391.

344. Fukumoto, M.; Estensen, R. D.; Sha, L.; et al. Association of Ki-ras with amplified DNA sequences, detected in human ovarian carcinomas by a modified in-gel renaturation assay. *Cancer Res.* **1989**, *49*, 1693–1697.

345. Zhou, D. J.; Gonzalez-Cadavid, N.; Ahuja, H.; Battifora, H.; Moore, G. E.; Cline, M. J. A unique pattern of proto-oncogene abnormalities in ovarian adenocarcinomas. *Cancer* **1988**, *62*, 1573–1576.

346. Schreiber, G.; Dubeau, L. C-*myc* proto-oncogene amplification detected by polymerase chain reaction in archival human ovarian carcinomas. *Am. J. Pathol.* **1990**, *137*, 653–658.

347. Kohler, M.; Janz, I.; Wintzer, H-O.; Wagner, E.; Bauknecht, T. The expression of EGF receptors, EGF-like factors and c-*myc* in ovarian and cervical carcinomas and their potential clinical significance. *Anticancer Res.* **1989**, *9*, 1537–1548.

348. Tashiro, H.; Miyazaki, K.; Okamura, H.; Iwai, A.; Fukumoto, M. c-*myc* over-expression in human primary ovarian tumours: its relevance to tumour progression. *Int. J. Cancer* **1992**, *50*, 828–833.

349. Bauknecht, T.; Angel, P.; Kohler, M.; et al. Gene structure and expression analysis of the epidermal growth factor receptor, transforming growth factor-alpha, *myc, jun*, and metallothionein in human ovarian carcinomas. *Cancer* **1993**, *71*, 419–429.

350. Niwa, O.; Enoki, Y.; Yokoro, K. Overexpression and amplification of the c-*myc* gene in mouse tumors induced by chemicals and radiations. *Jpn. J. Cancer Res.* **1989**, *80*, 212–218.

351. Bauknecht, T.; Birmelin, G.; Kommoss, F. Clinical significance of oncogenes and growth factors in ovarian carcinomas. *J. Steroid Biochem. Mol. Biol.* **1990**, *37*, 855–862.

352. Watson, J. V.; Curling, O. M.; Munn, C. F.; Hudson, C. N. Oncogene expression in ovarian cancer: a pilot study of c-*myc* oncoprotein in serous papillary ovarian cancer. *Gynecol. Oncol.* **1987**, *28*, 137–150.

353. Polacarz, S. V.; Hey, N. A.; Stephenson, T. J.; Hill, A. S. c-*myc* oncogene product P62^{c-myc} in ovarian mucinous neoplasms: immunohistochemical study correlated with malignancy. *J. Clin. Pathol.* **1989**, *42*, 148–152.

354. Weisz, A.; Bresciani, F. Estrogen induces expression of c-*fos* and c-*myc* protooncogenes in rat uterus. *Mol. Endocrinol.* **1988**, *2*, 816–824.

355. Murphy, L. J.; Murphy, L. C.; Friesen, H. G. Estrogen induction of N-*myc* and c-*myc* proto-oncogene expression in the rat uterus. *Endocrinol.* **1987**, *120*, 1882–1888.

356. Travers, M. T.; Knowler, J. T. Oestrogen-induced expression of oncogenes in the immature rat uterus. *FEBS Lett.* **1987**, *211*, 27–30.

357. Rempel, S. A.; Johnston, R. N. Steroid-induced cell proliferation in vivo is associated with increased c-*myc* proto-oncogene transcript abundance. *Development* **1988**, *104*, 87–95.

358. Fink, K. L.; Wieben, E. D.; Woloschak, G. E.; Spelsberg, T. C. Rapid regulation of c-*myc* protooncogene expression by progesterone in the avian oviduct. *Proc. Natl. Acad. Sci. USA* **1988**, *85*, 1796–1800.

359. Itkes, A. V.; Imamova, L. R.; Alexandrova, N. M.; Favorova, O. O.; Kisselev, L. L. Expression of c-*myc* gene in human ovary carcinoma cells treated with vanadate. *Exp. Cell Res.* **1990**, *188*, 169–171.

360. Rories, C.; Lau, C. K.; Fink, K.; Spelsberg, T. C. Rapid inhibition of c-*myc* gene expression by a glucocorticoid in the avian oviduct. *Mol. Endocrinol.* **1989**, *3*, 991–1001.

361. Somay, C.; Grunt, T. W.; Mannhalter, C.; Dittrich, C. Relationship of *myc* protein expression to the phenotype and to the growth potential of HOC-7 ovarian cancer cells. *Br. J. Cancer* **1992**, *66*, 93–98.

362. Darling, D.; Tavassoli, M.; Linskens, M. H. K.; Farzaneh, F. DMSO induced modulation of c-*myc* steady-state RNA levels in a variety of different cell lines. *Oncogene* **1989**, *4*, 175–179.

363. Yarden, A.; Kimchi, A. Tumor necrosis factor reduces c-*myc* expression and cooperates with interferon-γ in HeLa cells. *Science* **1986**, *234*, 1419–1421.

364. Fujimoto, M.; Sheridan, P. J.; Sharp, Z. D.; Weaker, F. J.; Kagan-Hallet, K. S.; Story, J. L. Proto-oncogene analysis in brain tumors. *J. Neurosurg.* **1989**, *70*, 910–915.

365. MacGregor, D. N.; Ziff, E. B. Elevated c-*myc* expression in childhood medulloblastomas. *Pediatric Res.* **1990**, *28*, 63–68.

366. Wasson, J. C.; Saylors, R. L. III.; Zeltzer, P.; et al. Oncogene amplification in pediatric brain tumors. *Cancer Res.* **1990**, *50*, 2987–2990.

367. Bigner, S. H.; Friedman, H. S.; Vogelstein, B.; Oakes, W. J.; Bigner, D. D. Amplification of the c-*myc* gene in human medulloblastoma cell lines and xenografts. *Cancer Res.* **1990**, *50*, 2347–2350.

368. Friedman, H. S.; Burger, P. C.; Bigner, S. H.; et al. Phenotypic and genotypic analysis of a human medulloblastoma cell line and transplantable xenograft (D341 Med) demonstrating amplification of c-*myc*. *Am. J. Pathol.* **1988**, *130*, 472–484.

369. Sawyer, J. R.; Swanson, C. M.; Roloson, G. J.; Longee, D. C.; Boop, F. A.; Chadduck, W. M. Molecular cytogenetic analysis of a medulloblastoma with isochromosome 17 and double minutes. *Cancer Genet. Cytogenet.* **1991**, *57*, 181–186.

370. Raffel, C.; Gilles, F. E.; Weinberg, K. I. Reduction to homozygosity and gene amplification in central nervous system primitive neuroectodermal tumors of childhood. *Cancer Res.* **1990**, *50*, 587–591.

371. Garson, J. A.; Pemberton, L. F.; Sheppard, P. W.; Varndell, I. M.; Coakham, H. B.; Kemshead, J. T. N-*myc* gene expression and oncoprotein characterisation in medulloblastoma. *Br. J. Cancer* **1989**, *59*, 889–894.

372. Trent, J.; Meltzer, P.; Rosenblum, M.; et al. Evidence for rearrangement, amplification, and expression of c-*myc* in a human glioblastoma. *Proc. Natl. Acad. Sci. USA* **1986**, *83*, 470–473.

373. Wong, A. J.; Bigner, S. H.; Bigner, D. D.; Kinzler, K. W.; Hamilton, S. R.; Vogelstein, B. Increased expression of the epidermal growth factor receptor gene in malignant gliomas is invariably associated with gene amplification. *Proc. Natl. Acad. Sci. USA* **1987**, *84*, 6899–6903.

374. Engelhard, H. H. III.; Butler, A. B. IV.; Bauer, K. D. Quantification of the c-*myc* oncoprotein in human glioblastoma cells and tumor tissue. *J. Neurosurg.* **1989**, *71*, 224–232.

375. Orian, J. M.; Vasilopoulos, K.; Yoshida, S.; Kaye, A. H.; Chow, C. W.; Gonzales, M. F. Overexpression of multiple oncogenes related to histological grade of astrocytic glioma. *Br. J. Cancer* **1992**, *66*, 106–112.

376. Tanaka, K.; Sato, C.; Maeda, Y.; et al. Establishment of a human malignant meningioma cell line with amplified c-*myc* oncogene. *Cancer* **1989**, *64*, 2243–2249.

377. Kazumoto, K.; Tamura, M.; Hoshino, H.; Yuasa, Y. Enhanced expression of the *sis* and c-*myc* oncogenes in human meningiomas. *J. Neurosurg.* **1990**, *72*, 786–791.

378. Xu, L.; Morgenbesser, S. D.; DePinho, R. A. Complex transcriptional regulation of *myc* family gene expression in the developing mouse brain and liver. *Mol. Cell. Biol.* **1991**, *11*, 6007–6015.

379. Hirvonen, H.; Mäkelä, T. P.; Sandberg, M.; Kalimo, H.; Vuorio, E.; Alitalo, K. Expression of the *myc* proto-oncogenes in developing human fetal brain. *Oncogene* **1990**, *5*, 1787–1797.

380. Ruppert, C.; Goldowitz, D.; Wille, W. Proto-oncogene c-*myc* is expressed in cerebellar neurons at different developmental stages. *EMBO J.* **1986**, *5*, 1897–1901.

381. Casalbore, P.; Agostini, E.; Alemà, S.; Falcone, G.; Tatò, F. The v-*myc* oncogene is sufficient to induce growth transformation of chick neuroretina cells. *Nature* **1987**, *326*, 188–190.

382. Fauquet, M.; Stehelin, D.; Saule, S. *myc* products induce the expression of catecholaminergic traits in quail neural crest-derived cells. *Proc. Natl. Acad. Sci. USA* **1990**, *87*, 1546–1550.

383. Blasi, E.; Barluzzi, R.; Bocchini, V.; Mazzolla, R.; Bistoni, F. Immortalization of murine microglial cells by a v-*raf*/v-*myc* carrying retrovirus. *J. Neuroimmunol.* **1990**, *27*, 229–237.

384. Bartlett, P. F.; Reid, H. H.; Bailey, K. A.; Bernard. O. Immortalization of mouse neural precursor cells by the c-*myc* oncogene. *Proc. Natl. Acad. Sci. USA* **1988**, *85*, 3255–3259.

385. Benvenisty, N.; Ornitz, D. M.; Bennett, G. L.; et al. Brain tumours and lymphomas in transgenic mice that carry HTLV-I LTR/c-*myc* and Ig/*tax* genes. *Oncogene* **1992**, *7*, 2399–2405.

386. Wiestler, O. D.; Aguzzi, A.; Schneemann, M.; Eibl, R.; von Deimling, A.; Kleihues, P. Oncogene complementation in fetal brain transplants. *Cancer Res.* **1992**, *52*, 3760–3767.

387. Fleming, W. H.; Hamel, A.; MacDonald, R.; et al. Expression of the c-*myc* protooncogene in human prostatic carcinoma and benign prostatic hyperplasia. *Cancer Res.* **1986**, *46*, 1535–1538.

388. Phillips, M. E. A.; Ferro, M. A.; Smith, P. J. B.; Davies, P. Intranuclear androgen receptor deployment and protooncogene expression in human diseased prostate. *Urol. Int.* **1987**, *42*, 115–119.

389. Matusik, R. J.; Fleming, W. H.; Hamel, A.; et al. Expression of the c-*myc* proto-oncogene in prostatic tissue. *Prog. Clin. Biol. Res.* **1987**, *239*, 91–112.

390. Fukumoto, M.; Shevrin, D. H.; Roninson, I. B. Analysis of gene amplification in human tumor cells. *Proc. Natl. Acad. Sci. USA* **1988**, *85*, 6846–6850.

391. Fox, S. B.; Persad, R. A.; Royds, J.; Kore, R. N.; Silcocks, P. B.; Collins, C. C. p53 and c-*myc* expression in stage A1 prostatic adenocarcinoma: useful prognostic determinants? *J. Urol.* **1993**, *150*, 490–494.

392. Thompson, T. C.; Southgate, J.; Kitchener, G.; Land, H. Multistage carcinogenesis induced by *ras* and *myc* oncogenes in a reconstituted organ. *Cell* **1989**, *56*, 917–930.

393. Thompson, T. C.; Timme, T. L.; Kadmon, D.; Park, S. H.; Egawa, S.; Yoshida, K. Genetic predisposition and mesenchymal-epithelial interactions in *ras* + *myc*-induced carcinogenesis in reconstituted mouse prostate. *Mol. Carcinogen.* **1993**, *7*, 165–179.

394. Lu, X.; Park, S. H.; Thompson, T. C.; Lane, D. P. *ras*-induced hyperplasia occurs with mutation of p53, but activated *ras* and *myc* together can induce carcinoma without p53 mutation. *Cell* **1992**, *70*, 153–161.

395. Katz, A. E.; Benson, M. C.; Wise, G. J. et al. Gene activity during the early phase of androgen-stimulated rat prostate regrowth. *Cancer Res.* **1989**, *49*, 5889–5894.

396. Wolf, D. A.; Kohlhuber, F.; Schulz, P.; Fittler, F.; Eick, D. Transcriptional down-regulation of c-*myc* in human prostate carcinoma cells by the synthetic androgen mibolerone. *Br. J. Cancer* **1992**, *65*, 376–382.

397. Egawa, S.; Kadmon, D.; Miller, G. J.; Scardino, P. T.; Thompson, T. C. Alterations in mRNA levels for growth-related genes after transplantation into castrated hosts in oncogene-induced clonal mouse prostate carcinoma. *Mol. Carcinogen.* **1992**, *5*, 52–61.

398. Thompson, T. C.; Egawa, S.; Kadmon, D.; et al. Androgen sensitivity and gene expression in *ras* + *myc*-induced mouse prostate carcinomas. *J. Steroid Biochem. Mol. Biol.* **1992**, *43*, 79–85.

399. Sikora, K.; Evan, G.; Stewart, J.; Watson, J. V. Detection of the c-*myc* oncogene product in testicular cancer. *Br. J. Cancer* **1985**, *52*, 171–176.

400. Watson, J. V.; Stewart, J.; Evan, G. I.; Ritson, A.; Sikora, K. The clinical significance of flow cytometric c-*myc* oncoprotein quantitation in testicular cancer. *Br. J. Cancer* **1986**, *53*, 331–337.

401. Stewart, T. A.; Bellvé, A. R.; Leder, P. Transcription and promoter usage of the *myc* gene in normal somatic and spermatogenic cells. *Science* **1984**, *226*, 707–710.

402. Lin, T.; Blaisdell, J.; Barbour, K. W.; Thompson, E. A. Transient activation of c-*myc* protooncogene expression in leydig cells by human chorionic gonadotropin. *Biochem. Biophys. Res. Commun.* **1988**, *157*, 121–126.

403. Czerwiec, F. S.; Melner, M. H.; Puett, D. Transiently elevated levels of c-*fos* and c-*myc* oncogene messenger ribonucleic acids in cultured murine leydig tumor cells after addition of human chorionic gonadotropin. *Mol. Endocrinol.* **1989**, *3*, 105–109.

404. Hall, S. H.; Berthelon, M-C.; Avallet, O.; Saez, J. M. Regulation of c-*fos*, c-*jun*, *jun*-B, and c-*myc* messenger ribonucleic acids by gonadotropin and growth factors in cultured pig leydig cell. *Endocrinol.* **1991**, *129*, 1243–1249.

405. Kinouchi, T.; Saiki, S.; Naoe, T.; et al. Correlation of c-*myc* expression with nuclear pleomorphism in human renal cell carcinoma. *Cancer Res.* **1989**, *49*, 3627–3630.

406. Yao, M.; Shuin, T.; Misaki, H.; Kubota, Y. Enhanced expression of c-*myc* and epidermal growth factor receptor (c-*erb*B-1) genes in primary human renal cancer. *Cancer Res.* **1988**, *48*, 6753–6757.

407. Slamon, D. J.; deKernion, J. B.; Verma, I. M.; Cline, M. J. Expression of cellular oncogenes in human malignancies. *Science* **1984**, *224*, 256–262.

408. Kubota, Y.; Shuin, T.; Yao, M.; Inoue, H.; Yoshioka, T. The enhanced ^{32}P labeling of CDP-diacyl-glycerol in c-*myc* gene expressed human kidney cancer cells. *FEBS Lett.* **1987**, *212*, 159–162.

409. Liehr, J. G.; Chiappetta, C.; Roy, D.; Stancel, G. M. Elevation of protooncogene messenger RNAs in estrogen-induced kidney tumors in the hamster. *Carcinogenesis* **1992**, *13*, 601–604.

410. Gemmill, R. M.; Coyle-Morris, J.; Ware-Uribe, L.; et al. A 1.5-megabase restriction map surrounding MYC does not include the translocation breakpoint in familial renal cell carcinoma. *Genomics* **1989**, *4*, 28–35.

411. Cowley, B. D. Jr.; Smardo, F. L.; Grantham, J. J.; Calvet, J. P. Elevated c-*myc* protooncogene expression in autosomal recessive polycystic kidney disease. *Proc. Natl. Acad. Sci. USA* **1987**, *84*, 8394–8398.

412. Cowley, B. D. Jr.; Chadwick, L. J.; Grantham, J. J.; Calvet, J. P. Sequential protooncogene expression in regenerating kidney following acute renal injury. *J. Biol. Chem.* **1989**, *264*, 8389–8393.

413. Mugrauer, G.; Ekblom, P. Contrasting expression patterns of three members of the *myc* family of protooncogenes in the developing and adult mouse kidney. *J. Cell Biol.* **1991**, *112*, 13–25.

414. Harding, M. A.; Gattone, V. H. II; Grantham, J. J.; Calvet, J. P. Localization of overexpressed c-*myc* mRNA in polycystic kidneys of the *cpk* mouse. *Kidney Intl.* **1992**, *41*, 317–325.

415. Trudel, M.; D'Agati, V.; Costantini, F. c-*myc* as an inducer of polycystic kidney disease in transgenic mice. *Kidney Intl.* **1991**, *39*, 665–671.

416. Beer, D. G.; Zweifel, K. A.; Simpson, D. P.; Pitot, H. C. Specific gene expression during compensatory renal hypertrophy in the rat. *J. Cell Physiol.* **1987**, *131*, 29–35.

417. Norman, J. T.; Bohman, R. E.; Fischmann, G.; et al. Patterns of mRNA expression during early cell growth differ in kidney epithelial cells destined to undergo compensatory hypertrophy versus regenerative hyperplasia. *Proc. Natl. Acad. Sci. USA* **1988**, *85*, 6768–6772.

418. Asselin, C.; Marcu, K. B. Mode of c-*myc* gene regulation in folic acid-induced kidney regeneration. *Oncogene Res.* **1989**, *5*, 67–72.

419. Murphy, L. J.; Bell, G. I.; Friesen, H. G. Growth hormone stimulates sequential induction of c-*myc* and insulin-like growth factor I expression *in vivo. Endocrinol.* **1987**, *120*, 1806–1812.

420. Wagner, H. E.; Steele, G. Jr.; Summerhayes, I. C. Preneoplastic lesions induced by *myc* and *src* oncogenes in reconstituted mouse bladder. *Surgery* **1990**, *108*, 146–153.

421. del Senno, L.; Maestri, I.; Piva, R.; et al. Differential hypomethylation of the c-*myc* protooncogene in bladder cancers at different stages and grades. *J. Urol.* **1989**, *142*, 146–149.

422. Debiec-Rychter, M.; Jones, R. F.; Zukowski, K.; Wang, C. Y. Oncogene expression of FANFT- or BBN-induced rat urothelial cells. *Int. J. Cancer* **1990**, *46*, 913–918.

423. Skouv, J.; Chistensen, B.; Autrup, H. Differential induction of transcription of c-*myc* and c-*fos* proto-oncogenes by 12-O-tetradecanoylphorbol-13-acetate in mortal and immortal human urothelial cells. *J. Cell Biochem.* **1987**, *34*, 71–79.

424. Stacey, S. N.; Nielsen, I.; Skouv, J.; Hansen, C.; Autrup, H. Deregulation in *trans* of c-*myc* expression in immortalized human urothelial cells and in T24 bladder carcinoma cells. *Mol. Carcinogen.* **1990**, *3*, 216–225.

425. Suarez, H. G.; Nardeux, P. C.; Andeol, Y.; Sarasin, A. Multiple activated oncogenes in human tumors. *Oncogene Res.* **1987**, *1*, 201–207.

426. Jian-Ren, G.; Li-Fu, H.; Yuan-Ching, C.; Da-Fong, W. Oncogenes in human primary hepatic cancer. *J. Cell Physiol.* **1986**, Supplement 4: 13–20.

427. Zhang, X-K.; Huang, D-P.; Chiu, D-K.; Chiu, J-F. The expression of oncogenes in human developing liver and hepatomas. *Biochem. Biophys. Res. Commun.* **1987**, *142*, 932–938.

428. Himeno, Y.; Fukuda, Y.; Hatanaka, M.; Imura, H. Expression of oncogenes in human liver disease. *Liver* **1988**, *8*, 208–212.

429. Arbuthnot, P.; Kew, M.; Fitschen, W. c-*fos* and c-*myc* oncoprotein expression in human hepato-cellular carcinomas. *Anticancer Res.* **1991**, *11*, 921–924.

430. Zhang, X-k.; Huang, D-p.; Qiu, D-K.; Chiu, J-f. The expression of c-*myc* and c-N-*ras* in human cirrhotic livers, hepatocellular carcinomas and liver tissue surrounding the tumors. *Oncogene* **1990**, *5*, 909–914.

431. Gan, F-Y.; Gesell, M. S.; Alousi, M.; Luk, G. D. Analysis of ODC and c-*myc* gene expression in hepatocellular carcinoma by in situ hybridization and immunohistochemistry. *J. Histochem. Cytochem.* **1993**, *41*, 1185–1196.

432. Su, T-S.; Lin, L-H.; Lui, W-Y.; et al. Expression of c-*myc* gene in human hepatoma. *Biochem. Biophys. Res. Commun.* **1985**, *132*, 264–268.

433. Hayashi, H.; Taira, M.; Tatibana, M.; Tabata, Y.; Isono, K. Elevation of c-*myc* transcript level in human liver during surgical resection of hepatocellular carcinoma: possible cause for underestimation of c-*myc* gene activation in the tumor. *Biochem. Biophys. Res. Commun.* **1989**, *162*, 1260–1264.

434. Tiniakos, D.; Spandidos, D. A.; Kakkanas, A.; Pintzas, A.; Pollice, L.; Tiniakos, G. Expression of *ras* and *myc* oncogenes in human hepatocellular carcinoma and non-neoplastic liver tissues. *Anticancer Res.* **1989**, *9*, 715–722.

435. Tiniakos, D.; Spandidos, D. A.; Yiagnisis, M.; Tiniakos, G. Expression of *ras* and c-*myc* oncoproteins and hepatitis B surface antigen in human liver disease. *Hepato. Gastroenterol.* **1993**, *40*, 37–40.

436. Huber, B. E.; Dearfield, K. L.; Williams, J. R.; Heilman, C. A.; Thorgeirsson, S. S. Tumorigenicity and transcriptional modulation of c-*myc* and N-*ras* oncogenes in a human hepatoma cell line. *Cancer Res.* **1985**, *45*, 4322–4329.

437. Huber, B. E.; Thorgeirsson, S. S. Analysis of c-*myc* expression in a human hepatoma cell line. *Cancer Res.* **1987**, *47*, 3414–3420.

438. Kaneko, Y.; Shibuya, M.; Nakayama, T.; et al. Hypomethylation of c-*myc* and epidermal growth factor receptor genes in human hepatocellular carcinoma and fetal liver. *Jpn. J. Cancer Res.* **1985**, *76*, 1136–1140.

439. Nambu, S.; Inoue, K.; Sasaki, H. Site-specific hypomethylation of the c-*myc* oncogene in human hepatocellular carcinoma. *Jpn. J. Cancer Res.* **1987**, *78*, 695–704.

440. Hayashi, K.; Makino, R.; Sugimura, T. Amplification and overexpression of the c-*myc* gene in Morris hepatomas. *Jpn. J. Cancer Res.* **1984**, *75*, 475–478.

441. Cote, G. J.; Chiu, J-F. The expressions of oncogenes and liver-specific genes in Morris hepatomas. *Biochem. Biophys. Res. Commun.* **1987**, *143*, 624–629.

442. Suchy, B. K.; Sarafoff, M.; Kerler, R.; Rabes, H. M. Amplification, rearrangements, and enhanced expression of c-*myc* in chemically induced rat liver tumors *in vivo* and *in vitro*. *Cancer Res.* **1989**, *49*, 6781–6787.

443. Tashiro, F.; Morimura, S.; Hayashi, K.; et al. Expression of the c-Ha-*ras* and c-*myc* genes in aflatoxin B_1-induced hepatocellular carcinomas. *Biochem. Biophys. Res. Commun.* **1986**, *138*, 858–864.

444. Fujimoto, Y.; Ishizaka, Y.; Tahira, T.; et al. Possible involvement of c-*myc* but not *ras* genes in hepatocellular carcinomas developing after spontaneous hepatitis in LEC rats. *Mol. Carcinogen.* **1991**, *4*, 269–274.

445. Chandar, N.; Lombardi, B.; Locker, J. c-*myc* gene amplification during hepatocarcinogenesis by a choline-devoid diet. *Proc. Natl. Acad. Sci. USA* **1989**, *86*, 2703–2707.

446. Makino, R.; Hayashi, K.; Sato, S.; Sugimura, T. Expressions of the c-Ha-*ras* and c-*myc* genes in rat liver tumors. *Biochem. Biophys. Res. Commun.* **1984**, *119*, 1096–1102.

447. Cote, G. J.; Lastra, B. A.; Cook, J. R.; Huang, D-P.; Chiu, J-F. Oncogene expression in rat hepatomas and during hepatocarcinogenesis. *Cancer Lett.* **1985**, *26*, 121–127.

448. Beer, D. G.; Schwarz, M.; Sawada, N.; Pitot, H. C. Expression of H-*ras* and c-*myc* protooncogenes in isolated γ-glutamyl transpeptidase-positive rat hepatocytes and in hepatocellular carcinomas induced by diethylnitrosamine. *Cancer Res.* **1986**, *46*, 2435–2441.

449. Hsieh, L. L.; Hsiao, W-L.; Peraino, C.; Maronpot, R. R.; Weinstein, I. B. Expression of retroviral sequences and oncogenes in rat liver tumors induced by diethylnitrosamine. *Cancer Res.* **1987**, *47*, 3421–3424.

450. Nagy, P.; Evarts, R. P.; Marsden, E.; Roach, J.; Thorgeirsson, S. S. Cellular distribution of c-*myc* transcripts during chemical hepatocarcinogenesis in rats. *Cancer Res.* **1988**, *48*, 5522–5527.

451. Huber, B. E.; Heilman, C. A.; Thorgeirsson, S. S. Poly(A⁺)RNA levels of growth-, differentiation- and transformation-associated genes in the progressive development of hepatocellular carcinoma in the rat. *Hepatology* **1989**, *9*, 756–762.

452. Porsch Hällström, I.; Gustafsson, J-Å.; Blanck, A. Effects of growth hormone on the expression of c-*myc* and c-*fos* during early stages of sex-differentiated rat liver carcinogenesis in the resistant hepatocyte model. *Carcinogenesis* **1989**, *10*, 2339–2343.

453. Porsch-Hällström, I.; Blanck, A.; Eriksson, L. C.; Gustafsson, J-Å. Expression of the c-*myc*, c-*fos* and c-*ras*^{Ha} protooncogenes during sex-differentiated rat liver carcinogenesis in the resistant hepatocyte model. *Carcinogenesis* **1989**, *10*, 1793–1800.

454. Porsch Hällström, I.; Gustafsson, J-Å.; Blanck, A. Role of growth hormone in the regulation of the c-*myc* gene during progression of sex-differentiated rat liver carcinogenesis in the resistant hepatocyte model. *Mol. Carcinogen.* **1991**, *4*, 376–381.

455. Yaswen, P.; Goyette, M.; Shank, P. R.; Fausto, N. Expression of c-Ki-*ras*, c-Ha-*ras*, and c-*myc* in specific cell types during hepatocarcinogenesis. *Mol. Cell. Biol.* **1985**, *5*, 780–786.

456. Braun, L.; Mikumo, R.; Fausto, N. Production of hepatocellular carcinoma by oval cells: cell cycle expression of c-*myc* and p53-at different stages of oval cell transformation. *Cancer Res.* **1989**, *49*, 1554–1561.

457. Tsao, M-S.; Sheperd, J.; Batist, G. Phenotypic expression in spontaneously transformed cultured rat liver epithelial cells. *Cancer Res.* **1990**, *50*, 1941–1947.

458. Huggett, A. C.; Ellis, P. A.; Ford, C. P.; Hampton, L. L.; Rimoldi, D.; Thorgeirsson, S. S. Development of resistance to the growth inhibitory effects of transforming growth factor β1 during the spontaneous transformation of rat liver epithelial cells. *Cancer Res.* **1991**, *51*, 5929–5936.

459. Beasley, R. P. Hepatitis B virus: the major etiology of hepatocellular carcinoma. *Cancer* **1988**, *61*, 1942–1956.

460. Möröy, T.; Marchio, A.; Etiemble, J.; Trépo, C.; Tiollais, P.; Buendia, M-A. Rearrangement and enhanced expression of c-*myc* in hepatocellular carcinoma of hepatitis virus infected wood-chucks. *Nature* **1986**, *324*, 276–279.

461. Hsu, T-y.; Möröy, T.; Etiemble, J.; et al. Activation of c-*myc* by woodchuck hepatitis virus insertion in hepatocellular carcinoma. *Cell*, **1988**, *55*, 627–635.

462. Etiemble, J.; Möröy, T.; Jacquemin, E.; Tiollais, P.; Buendia, M-A. Fused transcripts of c-*myc* and a new cellular locus, *hcr* in a primary liver tumor. *Oncogene* **1989**, *4*, 51–57.

463. Möröy, T.; Etiemble, J.; Bougueleret, L.; Hadchouel, M.; Tiollais, P.; Buendia, M-A. Structure and expression of *hcr*, a locus rearranged with c-*myc* in a woodchuck hepatocellular carcinoma. *Oncogene* **1989**, *4*, 59–65.

464. Fourel, G.; Trepo, C.; Bougueleret, L.; et al. Frequent activation of N-*myc* genes by hepadnavirus insertion in woodchuck liver tumours. *Nature* **1990**, *347*, 294–298.

465. Hansen, L. J.; Tennant, B. C.; Seeger, C.; Ganem, D. Differential activation of *myc* gene family members in hepatic carcinogenesis by closely related hepatitis B viruses. *Mol. Cell. Biol.* **1993**, *13*, 659–667.

466. Transy, C.; Fourel, G.; Robinson, W. S.; Tiollais, P.; Marion, P. L.; Buendia, M-A. Frequent amplification of c-*myc* in ground squirrel liver tumors associated with past or ongoing infection with a hepadnavirus. *Proc. Natl. Acad. Sci. USA* **1992**, *89*, 3874–3878.

467. Balsano, C.; Avantaggiati, M. L.; Natoli, G.; et al. Full-length and truncated versions of the hepatitis B virus (HBV) X protein (pX) transactivate the cMYC protooncogene at the transcriptional level. *Biochem. Biophys. Res. Commun.* **1991**, *176*, 985–992.

468. Sandgren, E. P.; Quaife, C. J.; Pinkert, C. A.; Palmiter, R. D.; Brinster, R. L. Oncogene-induced liver neoplasia in transgenic mice. *Oncogene* **1989**, *4*, 715–724.

469. Garfield, S.; Huber, B. E.; Nagy, P.; Cordingley, M. G.; Thorgeirsson, S. S. Neoplastic transformation and lineage switching of rat liver epithelial cells by retrovirus-associated oncogenes. *Mol. Carcinogen.* **1988**, *1*, 189–195.

470. Muakkassah-Kelly, S. F.; Jans, D. A.; Lydon, N.; et al. Electroporation of cultured adult rat hepatocytes with the c-*myc* gene potentiates DNA synthesis in response to epidermal growth factor. *Exp. Cell Res.* **1988**, *178*, 296–306.

471. Skouteris, G. G.; Kaser, M. R. Expression of exogenous c-*myc* oncogene does not initiate DNA synthesis in primary rat hepatocyte cultures. *J. Cell Physiol.* **1992**, *150*, 353–359.

472. Strom, S. C.; Faust, J. B.; Cappelluti, E.; Harris, R. B.; Lalwani, N. D. Characterization of liver epithelial cells transfected with *myc* and/or *ras* oncogenes. *Dig. Dis. Sci.* **1991**, *36*, 642–652.

473. Lee, L. W.; Raymond, V. W.; Tsao, M-S.; Lee, D. C.; Earp, H. S.; Grisham, J. W. Clonal cosegregation of tumorigenicity with overexpression of c-*myc* and transforming growth factor α genes in chemically transformed rat liver epithelial cells. *Cancer Res.* **1991**, *51*, 5238–5244.

474. Murakami, H.; Sanderson, N. D.; Nagy, P.; Marino, P. A.; Merlino, G.; Thorgeirsson, S. S. Transgenic mouse model for synergistic effects of nuclear oncogenes and growth factors in tumorigenesis: Interaction of c-*myc* and transforming growth factor α in hepatic oncogenesis. *Cancer Res.* **1993**, *53*, 1719–1723.

475. Sinha, S.; Neal, G. E.; Legg, R. F.; Watson, J. V.; Pearson, C. The expression of c-*myc* related to the proliferation and transformation of rat liver-derived epithelial cells. *Br. J. Cancer* **1989**, *59*, 674–676.

476. Morello, D.; Lavenu, A.; Babinet, C. Differential regulation and expression of jun, c-*fos* and c-*myc* proto-oncogenes during mouse liver regeneration and after inhibition of protein synthesis. *Oncogene* **1990**, *5*, 1511–1519.

477. Sobczak, J.; Mechti, N.; Tournier, M-F.; Blanchard, J-M.; Duguet, M. c-*myc* and c-*fos* gene regulation during mouse liver regeneration. *Oncogene* **1989**, *4*, 1503–1508.

478. Morello, D.; Fitzgerald, M. J.; Babinet, C.; Fausto, N. c-*myc*, c-*fos*, and c-jun regulation in the regenerating livers of normal and H-2K/c-*myc* transgenic mice. *Mol. Cell. Biol.* **1990**, *10*, 3185–3193.

479. Horikawa, S.; Sakata, K.; Uchiumi, F.; Hatanaka, M.; Tsukada, K. Effects of actinomycin D on DNA replication and c-*myc* expression during rat liver regeneration. *Biochem. Biophys. Res. Commun.* **1987**, *144*, 1049–1054.

480. Makino, R.; Hayashi, K.; Sugimura, T. c-*myc* transcript is induced in rat liver at a very early stage of regeneration or by cycloheximide treatment. *Nature* **1984**, *310*, 697–698.

481. Goyette, M.; Petropoulos, C. J.; Shank, P. R.; Fausto, N. Regulated transcription of c-Ki-*ras* and c-*myc* during compensatory growth of rat liver. *Mol. Cell. Biol.* **1984**, *4*, 1493–1498.

482. Fausto, N.; Shank, P. R. Oncogene expression in liver regeneration and hepatocarcinogenesis. *Hepatology* **1983**, *3*, 1016–1023.

483. Thompson, N. L.; Mead, J. E.; Braun, L.; Goyette, M.; Shank, P. R.; Fausto, N. Sequential protooncogene expression during rat liver regeneration. *Cancer Res.* **1986**, *46*, 3111–3117.

484. Sobczak, J.; Tournier, M-F.; Lotti, A-M.; Duguet, M. Gene expression in regenerating liver in relation to cell proliferation and stress. *Eur. J. Biochem.* **1989**, *180*, 49–53.

485. Messina, J. L. Inhibition and stimulation of c-*myc* gene transcription by insulin in rat hepatoma cells. *J. Biol. Chem.* **1991**, *266*, 17995–18001.

486. Taub, R.; Roy, A.; Dieter, R.; Koontz, J. Insulin as a growth factor in rat hepatoma cells. *J. Biol. Chem.* **1987**, *262*, 10893–10897.

487. Skouteris, G. G.; Kaser, M. R. Prostaglandins E_2 and F_{2a} mediate the increase in c-*myc* expression induced by EGF in primary rat hepatocyte cultures. *Biochem. Biophys. Res. Commun.* **1991**, *178*, 1240–1246.

488. Sawada, N. Hepatocytes from old rats retain responsiveness of c-*myc* expression to EGF in primary culture but do not enter S phase. *Exp. Cell Res.* **1989**, *181*, 584–588.

489. Skouteris, G. G.; McMenamin, M. Transforming growth factor-α-induced DNA synthesis and c-*myc* expression in primary rat hepatocyte cultures is modulated by indomethacin. *Biochem. J.* **1992**, *281*, 729–733.

490. Etienne, P. L.; Baffet, G.; Desvergne, B.; Boisnard-Rissel, M.; Glaise, D.; Guguen-Guillouzo, C. Transient expression of c-*fos* and constant expression of c-*myc* in freshly isolated and cultured normal adult rat hepatocytes. *Oncogene Res.* **1988**, *3*, 255–262.

491. Kumatori, A.; Nakamura, T.; Ichihara, A. Cell-density dependent expression of the c-*myc* gene in primary cultured rat hepatocytes. *Biochem. Biophys. Res. Commun.* **1991**, *178*, 480–485.

492. Vasudevan, S.; Lee, G.; Rao, P. M.; Rajalakshmi, S.; Sarma, D. S. R. Rapid and transient induction of c-*fos*, c-*myc* and c-Ha-*ras* in rat liver following glycine administration. *Biochem. Biophys. Res.* **1988**, *152*, 252–256.

493. Horikawa, S.; Sakata, K.; Hatanaka, M.; Tsukada, K. Expression of c-*myc* oncogene in rat liver by a dietary manipulation. *Biochem. Biophys. Res. Commun.* **1986**, *140*, 574–580.

494. Tichonicky, L.; Kruh, J.; Defer, N. Sodium butyrate inhibits c-*myc* and stimulates c-*fos* expression in all the steps of the cell-cycle in hepatoma tissue cultured cells. *Biol. Cell* **1990**, *69*, 65–67.

495. Kaneko, Y.; Toda, G.; Oka, H. Effects of teleocidin on the morphology and c-*myc* expression of hepatoma cells which are not inhibited by protein kinase antagonists. *Biochem. Biophys. Res. Commun.* **1987**, *145*, 549–555.

496. Porsch Hallstrom, I.; Gustafsson, J-A.; Blanck, A. Hypothalamo-pituitary regulation of the c-*myc* gene in rat liver. *J. Mol. Endocrinol.* **1990**, *5*, 267–274.

497. Yamada, H.; Yoshida, T.; Sakamoto, H.; Terada, M.; Sugimura, T. Establishment of a human pancreatic adenocarcinoma cell line (PSN-1) with amplifications of both c-*myc* and activated c-Ki-*ras* by a point mutation. *Biochem. Biophys. Res. Commun.* **1986**, *140*, 167–173.

498. Silverman, J. A.; Kuhlmann, E. T.; Zurlo, J.; Yager, J. D.; Longnecker, D. S. Expression of c-*myc*, c-*raf*-1, and c-Ki-*ras* in azaserine-induced pancreatic carcinomas and growing pancreas in rats. *Mol. Carcinogen.* **1990**, *3*, 379–386.

499. Hofler, H.; Ruhri, C.; Putz, B.; Wirnsberger, G.; Hauser, H. Oncogene expression in endocrine pancreatic tumors. *Virchows. Archiv. B Cell Pathol.* **1988**, *55*, 355–361.

500. Welsh, M.; Welsh, N.; Nilsson, T.; et al. Stimulation of pancreatic islet beta-cell replication by oncogenes. *Proc. Natl. Acad. Sci. USA* **1988**, *85*, 116–120.

501. Fredrickson, T. N.; Hartley, J. W.; Wolford, N. K.; Resau, J. H.; Rapp, U. R.; Morse, H. C. III. Histogenesis and clonality of pancreatic tumors induced by v-*myc* and v-*raf* oncogenes in NFS/N mice. *Am. J. Pathol.* **1988**, *131*, 444–451.

502. Quaife, C. J.; Pinkert, C. A.; Ornitz, D. M.; Palmiter, R. D.; Brinster, R. L. Pancreatic neoplasia induced by *ras* expression in acinar cells of transgenic mice. *Cell* **1987**, *48*, 1023–1034.

503. Calvo, E. L.; Dusetti, N. J.; Cadenas, M. B.; Dagorn, J-C.; Iovanna, J. L. Changes in gene expression during pancreatic regeneration: Activation of c-*myc* and H-*ras* oncogenes in the rat pancreas. *Pancreas* **1991**, *6*, 150–156.

504. Lu, L.; Logsdon, C. D. CCK, bombesin, and carbachol stimulate c-*fos*, c-jun, and c-*myc* oncogene expression in rat pancreatic acini. *Am. J. Physiol.* **1992**, *263*, G327–G332.

505. Tsuboi, K.; Hirayoshi, K.; Takeuchi, K.; et al. Expression of the c-*myc* gene in human gastrointestinal malignancies. *Biochem. Biophys. Res. Commun.* **1987**, *146*, 699–704.

506. Lu, S-H.; Hsieh, L-L.; Luo, F-C.; Weinstein, I. B. Amplification of the EGF receptor and c-*myc* genes in human esophageal cancers. *Int. J. Cancer* **1988**, *42*, 502–505.

507. Miyazaki, S.; Sasno, H.; Shiga, K.; et al. Analysis of c-*myc* oncogene in human esophageal carcinoma: immunohistochemistry, *in situ* hybridization and Northern and Southern blot studies. *Anticancer Res.* **1992**, *12*, 1747–1756.

508. Allum, W.H.; Newbold, K.M.; Macdonald, F.; Russell, B.; Stokes, H. Evaluation of p62^{c-myc} in benign and malignant gastric epithelia. *Br. J. Cancer* **1987**, *56*, 785–786.

509. Yamamoto, T.; Yasui, W.; Ochiai, A.; et al. Immunohistochemical detection of c-*myc* oncogene product in human gastric carcinomas: expression in tumor cells and stromal cells. *Jpn. J. Cancer Res.* **1987**, *78*, 1169–1174.

510. Ciclitira, P.J.; Macartney, J.C.; Evan, G. Expression of c-*myc* in non-malignant and pre-malignant gastrointestinal disorders. *J. Pathol.* **1987**, *151*, 293–296.

511. Karayiannis, M.; Yiagnisis, M.; Papadimitriou, K.; Field, J.K.; Spandidos, D.A. Evaluation of the *ras* and *myc* oncoproteins in benign gastric lesions. *Anticancer Res.* **1990**, *10*, 1127–1134.

512. Koda, T.; Matsushima, S.; Sasaki, A.; Danjo, Y.; Kakinuma, M. c-*myc* gene amplification in primary stomach cancer. *Jpn. J. Cancer Res.* **1985**, *76*, 551–554.

513. Nakasato, F.; Sakamoto, H.; Mori, M.; et al. Amplification of the c-*myc* oncogene in human stomach cancers. *Jpn. J. Cancer Res.* **1984**, *75*, 737–742.

514. Shibuya, M.; Yokota, J.; Ueyama, Y. Amplification and expression of a cellular oncogene (c-*myc*) in human gastric adenocarcinoma cells. *Mol. Cell. Biol.* **1985**, *5*, 414–418.

515. Kim, S. W.; Beauchamp, R. D.; Townsend, C. M. Jr.; Thompson, J. C. Vasoactive intestinal polypeptide inhibits c-*myc* expression and growth of human gastric carcinoma cells. *Surgery* **1991**, *110*, 270–276.

516. Yoshida, K.; Takanashi, A.; Kyo, E.; et al. Epidermal growth factor induces the expression of its receptor gene in human gastric carcinoma cell line TMK-1. *Jpn. J. Cancer Res.* **1989**, *80*, 743–746.

517. Erisman, M. D.; Rothberg, P. G.; Diehl, R. E.; Morse, C. C.; Spandorfer, J. M.; Astrin, S. M. Deregulation of c-*myc* gene expression in human colon carcinoma is not accompanied by amplification or rearrangement of the gene. *Mol. Cell. Biol.* **1985**, *5*, 1969–1976.

518. Guillem, J. G.; Levy, M. F.; Hsieh, L. L.; et al. Increased levels of phorbin, c-*myc*, and ornithine decarboxylase RNAs in human colon cancer. *Mol. Carcinogen.* **1990**, *3*, 68–74.

519. Mariani-Costantini, R.; Theillet, C.; Hutzell, P.; Merlo, G.; Schlom, J.; Callahan, R. In situ detection of c-*myc* mRNA in adenocarcinomas, adenomas, and mucosa of human colon. *J. Histochem. Cytochem.* **1989**, *37*, 293–298.

520. Monnat, M.; Tardy, S.; Saraga, P.; Diggelmann, H.; Costa, J. Prognostic implications of expression of the cellular genes *myc*, *fos*, Ha-*ras* and Ki-*ras* in colon carcinoma. *Int. J. Cancer* **1987**, *40*, 293–299.

521. Sikora, K.; Chan, S.; Evan, G.; et al. c-*myc* oncogene expression in colorectal cancer. *Cancer* **1987**, *59*, 1289–1295.

522. Untawale, S.; Blick, M. Oncogene expression in adenocarcinomas of the colon and in colon tumor-derived cell lines. *Anticancer Res.* **1988**, *8*, 1–8.

523. Klimpfinger, M.; Zisser, G.; Ruhri, C.; Pütz, B.; Steindorfer, P.; Hofler, H. Expression of c-*myc* and c-*fos* mRNA in colorectal carcinoma in man. *Virchows Archiv. B Cell Pathol.* **1990**, *59*, 165–171.

524. Matsumura, T.; Dohi, K.; Takanashi, A.; Ito, H.; Tahara, E. Alteration and enhanced expression of the c-*myc* oncogene in human colorectal carcinomas. *Path. Res. Pract.* **1990**, *186*, 205–211.

525. Finley, G. G.; Schulz, N. T.; Hill, S. A.; Geiser, J. R.; Pipas, J. M.; Meisler, A. I. Expression of the *myc* gene family in different stages of human colorectal cancer. *Oncogene* **1989**, *4*, 963–971.

526. Sugio, K.; Kurata, S.; Sasaki, M.; Soejima, J.; Sasazuki, T. Differential expression of c-*myc* gene and c-*fos* gene in premalignant and malignant tissues from patients with familial polyposis coli. *Cancer Res.* **1988**, *48*, 4855–4861.

527. Calabretta, B.; Kaczmarek, L.; Ming, P-M.L.; Au, F.; Ming, S-C. Expression of c-*myc* and other cell cycle-dependent genes in human colon neoplasia. *Cancer Res.* **1985**, *45*, 6000–6004.

528. Camplejohn, R. S. Cell Kinetics. *Recent Results Cancer Res.* **1982**, *83*, 21–30.

529. Ota, D. M.; Drewinko, B. Growth kinetics of human colorectal carcinoma. *Cancer Res.* **1985**, *45*, 2128–2131.

530. Camplejohn, R. S.; Bone, G.; Aherne, W. Cell proliferation in rectal carcinoma and rectal mucosa. A stathmokinetic study. *Eur. J. Cancer* **1973**, *9*, 577–581.

531. Bleiberg, H.; Salhadin, A.; Galand, P. Cell cycle parameters in human colon. *Cancer* **1977**, *39*, 1190–1194.

532. Detke, S.; Lichtler, A.; Phillips, I.; Stein, J.; Stein, G. Reassessment of histone gene expression during cell cycle in human cells by using homologous H4 histone cDNA. *Proc. Natl. Acad. Sci. USA* **1979**, *76*, 4995–4999.

533. Hann, S. R.; Thompson, C. B.; Eisenman, R. N. c-*myc* oncogene protein synthesis is independent of the cell cycle in human and avian cells. *Nature* **1985**, *314*, 366–369.

534. Thompson, C. B.; Challoner, P. B.; Neiman, P. E.; Groudine, M. Levels of c-*myc* oncogene mRNA are invariant throughout the cell cycle. *Nature* **1985**, *314*, 363–366.

535. Rabbitts, P. H.; Watson, J. V.; Lamond, A.; et al. Metabolism of c-*myc* gene products: c-*myc* mRNA and protein expression in the cell cycle. *EMBO J.* **1985**, *4*, 2009–2015.

536. Viel, A.; Maestro, R.; Toffoli, G.; Grion, G.; Boiocchi, M. c-*myc* overexpression is a tumor-specific phenomenon in a subset of human colorectal carcinomas. *J. Cancer Res. Clin. Oncol.* **1990**, *116*, 288–294.

537. Guillem, J. G.; Hsieh, L. L.; O'Toole, K. M.; Forde, K. A.; LoGerfo, P.; Weinstein, I. B. Changes in expression of oncogenes and endogenous retroviral-like sequences during colon carcinogenesis. *Cancer Res.* **1988**, *48*, 3964–3971.

538. Melhem, M. F.; Meisler, A. I.; Finley, G. G.; et al. Distribution of cells expressing *myc* proteins in human colorectal epithelium, polyps, and malignant tumors. *Cancer Res.* **1992**, *52*, 5853–5864.

539. Stewart, J.; Evan, G.; Watson, J.; Sikora, K. Detection of the c-*myc* oncogene product in colonic polyps and carcinomas. *Br. J. Cancer* **1986**, *53*, 1–6.

540. Jones, D. J.; Ghosh, A. K.; Moore, M.; Schofield, P. F. A critical appraisal of the immunohistochemical detection of the c-*myc* oncogene product in colorectal cancer. *Br. J. Cancer* **1987**, *56*, 779–783.

541. Tulchin, N.; Ornstein, L.; Harpaz, N.; Guillem, J.; Borner, C.; O'Toole, K. C-*myc* protein distribution. Neoplastic tissues of the human colon. *Am. J. Pathol.* **1992**, *140*, 719–729.

542. Pavelic, Z. P.; Pavelic, L.; Kuvelkar, R.; Gapany, S. R. High c-*myc* protein expression in benign colorectal lesions correlates with the degree of dysplasia. *Anticancer Res.* **1992**, *12*, 171–176.

543. Williams, A. R. W.; Piris, J.; Wyllie, A. H. Immunohistochemical demonstration of altered intracellular localization of the c-*myc* oncogene product in human colorectal neoplasms. *J. Pathol.* **1990**, *160*, 287–293.

544. Sundaresan, V.; Forgacs, I. C.; Wight, D. G. D.; Wilson, B.; Evan, G. I.; Watson, J. V. Abnormal distribution of c-*myc* oncogene product in familial adenomatous polyposis. *J. Clin. Pathol.* **1987**, *40*, 1274–1281.

545. Agnantis, N. J.; Apostolikas, N.; Sficas, C.; Zolota, V.; Spandidos, D. A. Immunohistochemical detection of *ras* p21 and c-*myc* p62 in colonic adenomas and carcinomas. *Hepato-gastroenterol.* **1991**, *38*, 239–242.

546. Erisman, M. D.; Scott, J. K.; Watt, R. A.; Astrin, S. M. The c-*myc* protein is constitutively expressed at elevated levels in colorectal carcinoma cell lines. *Oncogene* **1988**, *2*, 367–378.

547. Dolcetti, R.; De Re, V.; Viel, A.; Pistello, M.; Tavian, M.; Boiocchi, M. Nuclear oncogene amplification or rearrangement is not involved in human colorectal malignancies. *Eur. J. Cancer Clin. Oncol.* **1988**, *24*, 1321–1328.

548. Alexander, R. J.; Buxbaum, J. N.; Raicht, R. F. Oncogene alterations in primary human colon tumors. *Gastroenterol.* **1986**, *91*, 1503–1510.

549. Meltzer, S. J.; Ahnen, D. J.; Battifora, H.; Yokota, J.; Cline, M. J. Protooncogene abnormalities in colon cancers and adenomatous polyps. *Gastroenterol.* **1987**, *92*, 1174–1180.

550. Heerdt, B. G.; Molinas, S.; Deitch, D.; Augenlicht, L. H. Aggressive subtypes of human colorectal tumors frequently exhibit amplification of the c-*myc* gene. *Oncogene* **1991**, *6*, 125–129.

551. Erisman, M. D.; Litwin, S.; Keidan, R. D.; Comis, R. L.; Astrin, S. M. Noncorrelation of the expression of the c-*myc* oncogene in colorectal carcinoma with recurrence of disease or patient survival. *Cancer Res.* **1988**, *48*, 1350–1355.

552. Forgue-Lafitte, M-E.; Coudray, A-M.; Bréant, B.; Mešter, J. Proliferation of the human colon carcinoma cell line HT29: Autocrine growth and deregulated expression of the c-*myc* oncogene. *Cancer Res.* **1989**, *49*, 6566–6571.

553. Herold, K. M.; Rothberg, P. G. Amplification and activation of the c-*myc* oncogene in adenocarcinoma of the large bowel. In: *Familial Adenomatous Polyposis* (Herrera, L., Ed.) Alan R. Liss, New York, 1990, pp. 361–369.

554. Alitalo, K.; Schwab, M.; Lin, C. C.; Varmus, H. E.; Bishop, J. M. Homogeneously staining chromosomal regions contain amplified copies of an abundantly expressed cellular oncogene (c-*myc*) in malignant neuroendocrine cells from a human colon carcinoma. *Proc. Natl. Acad. Sci. USA* **1983**, *80*, 1707–1711.

555. La Rocca, R. V.; Park, J-G.; Danesi, R.; Del Tacca, M.; Steinberg, S. M.; Gazdar, A. F. Pattern of growth factor, proto-oncogene and carcinoembryonic antigen gene expression in human colorectal carcinoma cell lines. *Oncology* **1992**, *49*, 209–214.

556. Yander, G.; Halsey, H.; Kenna, M.; Augenlicht, L. H. Amplification and elevated expression of c-*myc* in a chemically induced mouse colon tumor. *Cancer Res.* **1985**, *45*, 4433–4438.

557. Schwab, M.; Klempnauer, K-H.; Alitalo, K.; Varmus, H.; Bishop, M. Rearrangement at the 5′ end of amplified c-*myc* in human COLO 320 cells is associated with abnormal transcription. *Mol. Cell. Biol.* **1986**, *6*, 2752–2755.

558. Cesarman, E.; Dalla-Favera, R.; Bentley, D.; Groudine, M. Mutations in the first exon are associated with altered transcription of c-*myc* in Burkitt lymphoma. *Science* **1987**, *238*, 1272–1275.

559. Zajac-Kaye, M.; Gelmann, E. P.; Levens, D. A point mutation in the c-*myc* locus of a Burkitt lymphoma abolishes binding of a nuclear protein. *Science* **1988**, *240*, 1776–1780.

560. Heruth, D. P.; Zirnstein, G. W.; Bradley, J. F.; Rothberg, P. G. Sodium butyrate causes an increase in the block to transcriptional elongation in the c-*myc* gene in SW837 rectal carcinoma cells. *J. Biol. Chem.* **1993**, *268*, 20466–20472.

561. Cedar, H. DNA methylation and gene activity. *Cell* **1988**, *53*, 3–4.

562. Sharrard, R. M.; Royds, J. A.; Rogers, S.; Shorthouse, A. J. Patterns of methylation of the c-*myc* gene in human colorectal cancer progression. *Br. J. Cancer* **1992**, *65*, 667–672.

563. Rothberg, P. G.; Spandorfer, J. M.; Erisman, M. D.; et al. Evidence that c-*myc* expression defines two genetically distinct forms of colorectal adenocarcinoma. *Br. J. Cancer* **1985**, *52*, 629–632.

564. Erisman, M. D.; Scott, J. K.; Astrin, S. M. Evidence that the familial adenomatous polyposis gene is involved in a subset of colon cancers with a complementable defect in c-*myc* regulation. *Proc. Natl. Acad. Sci. USA* **1989**, *86*, 4264–4268.

565. Rodriguez-Alfageme, C.; Stanbridge, E. J.; Astrin, S. M. Suppression of deregulated c-*myc* expression in human colon carcinoma cells by chromosome 5 transfer. *Proc. Natl. Acad. Sci. USA* **1992**, *89*, 1482–1486.

566. Bodmer, W. F.; Bailey, C. J.; Bodmer, J.; et al. Localization of the gene for familial adenomatous polyposis on chromosome 5. *Nature* **1987**, *328*, 614–616.

567. Tanaka, K.; Oshimura, M.; Kikuchi, R.; Seki, M.; Hayashi, T.; Miyaki, M. Suppression of tumorigenicity in human colon carcinoma cells by introduction of normal chromosome 5 or 18. *Nature* **1991**, *349*, 340–342.

568. Zhao, J.; Buick, R. N. Relationship of levels and kinetics of H-*ras* expression to transformed phenotype and loss of TGF-β1-mediated growth regulation in intestinal epithelial cells. *Exp. Cell Res.* **1993**, *204*, 82–87.

569. Shirasawa, S.; Furuse, M.; Yokoyama, N.; Sasazuki, T. Altered growth of human colon cancer cell lines disrupted at activated Ki-*ras*. *Science* **1993**, *260*, 85–88.

570. Pories, S. E.; Summerhayes, I. C.; Steele, G. D. Oncogene-mediated transformation. *Arch. Surg.* **1991**, *126*, 1387–1389.

571. Pories, S.; Jaros, K.; Steele, G. Jr.; Pauley, A.; Summerhayes, I. C. Oncogene-mediated transformation of fetal rat colon in vitro. *Oncogene* **1992**, *7*, 885–893.

572. D'Emilia, J. C.; Mathey-Prevot, B.; Jaros, K.; Wolf, B.; Steele, G. Jr.; Summerhayes, I. C. Preneoplastic lesions induced by *myc* and *src* oncogenes in a heterotopic rat colon. *Oncogene* **1991**, *6*, 303–309.

573. Niles, R. M.; Wilhelm, S. A.; Thomas, P.; Zamcheck, N. The effect of sodium butyrate and retinoic acid on growth and CEA production in a series of human colorectal tumor cell lines representing different states of differentiation. *Cancer Invest.* **1988**, *6*, 39–45.

574. Chung, Y. S.; Song, I. S.; Erickson, R. H.; Sleisenger, M. H.; Kim, Y. S. Effect of growth and sodium butyrate on brush border membrane-associated hydrolases in human colorectal cancer cell lines. *Cancer Res.* **1985**, *45*, 2976–2982.

575. Kim, Y. S.; Tsao, D.; Siddiqui, B.; et al. Effects of sodium butyrate and dimethylsulfoxide on biochemical properties of human colon cancer cells. *Cancer* **1980**, *45*, 1185–1192.

576. Gum, J. R.; Kam, W. K.; Byrd, J. C.; Hicks, J. W.; Sleisenger, M. H.; Kim, Y. S. Effects of sodium butyrate on human colonic adenocarcinoma cells. *J. Biol. Chem.* **1987**, *262*, 1092–1097.

577. Barnard, J. A.; Warwick, G. Butyrate rapidly induces growth inhibition and differentation in HT-29 cells. *Cell Growth & Differ.* **1993**, *4*, 495–501.

578. Tsao, D.; Shi, Z.; Wong, A.; Kim, Y. S. Effect of sodium butyrate on carcinoembryonic antigen production by human colonic adenocarcinoma cells in culture. *Cancer Res.* **1983**, *43*, 1217–1222.

579. Deng, G.; Liu, G.; Hu, L.; Gum, J. R. Jr.; Kim, Y. S. Transcriptional regulation of the human placental-like alkaline phosphatase gene and mechanisms involved in its induction by sodium butyrate. *Cancer Res.* **1992**, *52*, 3378–3383.

580. Herz, F.; Schermer, A.; Halwer, M.; Bogart, L. H. Alkaline phosphatase in HT-29, a human colon cancer cell line: influence of sodium butyrate and hyperosmolality. *Arch. Biochem.* **1981**, *210*, 581–591.

581. Taylor, C. W.; Kim, Y. S.; Childress-Fields, K. E.; Yeoman, L. C. Sensitivity of nuclear c-*myc* levels and induction to differentiation-inducing agents in human colon tumor cell lines. *Cancer Lett.* **1992**, *62*, 95–105.

582. Souleimani, A.; Asselin, C. Regulation of c-*myc* expression by sodium butyrate in the colon carcinoma cell line Caco-2. *FEBS Lett.* **1993**, *326*, 45–50.

583. Celano, P.; Baylin, S. B.; Giardiello, F. M.; Nelkin, B. D.; Casero, R. A. Jr. Effect of polyamine depletion on c-*myc* expression in human colon carcinoma cells. *J. Biol. Chem.* **1988**, *263*, 5491–5494.

584. Celano, P.; Baylin, S. B.; Casero, R. A. Jr. Polyamines differentially modulate the transcription of growth-associated genes in human colon carcinoma cells. *J. Biol. Chem.* **1989**, *264*, 8922–8927.

585. Chatterjee, D.; Savarese, T. M. Posttranscriptional regulation of c-*myc* proto-oncogene expression and growth inhibition by recombinant human interferon-β ser[17] in a human colon carcinoma cell line. *Cancer Chemotherapy Pharmacol.* **1992**, *30*, 12–20.

586. Mulder, K. M.; Levine, A. E.; Hernandez, X.; McKnight, M. K.; Brattain, D. E.; Brattain, M. G. Modulation of c-*myc* by transforming growth factor-β in human colon carcinoma cells. *Biochem. Biophys. Res. Commun.* **1988**, *150*, 711–716.

587. Brattain, M. G.; Levine, A. E.; Chakrabarty, S.; Yeoman, L. C.; Willson, J. K. V.; Long, B. H. Heterogeneity of human colon carcinoma. *Cancer Metastasis Rev.* **1984**, *3*, 177–191.

588. Mulder, K. M. Differential regulation of c-*myc* and transforming growth factor-α messenger RNA expression in poorly differentiated and well-differentiated colon carcinoma cells during the establishment of a quiescent state. *Cancer Res.* **1991**, *51*, 2256–2262.

589. Mulder, K. M.; Humphrey, L. E.; Choi, H. G.; Childress-Fields, K. E.; Brattain, M. G. Evidence for c-*myc* in the signaling pathway for TGF-β in well-differentiated human colon carcinoma cells. *J. Cell Physiol.* **1990**, *145*, 501–507.

590. Mulder, K. M.; Brattain, M. G. Effects of growth stimulatory factors on mitogenicity and c-*myc* expression in poorly differentiated and well differentiated human colon carcinoma cells. *Mol. Endocrinol.* **1989**, *3*, 1215–1222.

591. Mulder, K. M.; Zhong, Q.; Choi, H. G.; Humphrey, L. E.; Brattain, M. G. Inhibitory effects of transforming growth factor β1 on mitogenic response, transforming growth factor α, and c-*myc* in quiescent, well-differentiated colon carcinoma cells. *Cancer Res.* **1990**, *50*, 7581–7586.

592. Stopera, S. A.; Bird, R. P. Effects of all-*trans* retinoic acid as a potential chemopreventive agent on the formation of azoxymethane-induced aberrant crypt foci: differential expression of c-*myc* and c-*fos* mRNA and protein. *Int. J. Cancer* **1993**, *53*, 798–804.

593. Narayan, S.; Rajakumar, G.; Prouix, H.; Singh, P. Estradiol is trophic for colon cancer in mice: effect on ornithine decarboxylase and c-*myc* messenger RNA. *Gastroenterol.* **1992**, *103*, 1823–1832.

594. St. Clair, W. H.; St. Clair, D. K. Effect of the Bowman-Birk protease inhibitor on the expression of oncogenes in the irradiated rat colon. *Cancer Res.* **1991**, *51*, 4539–4543.

595. St. Clair, W. H.; Billings, P. C.; Kennedy, A. R. The effects of the Bowman-Birk protease inhibitor on c-*myc* expression and cell proliferation in the unirradiated and irradiated mouse colon. *Cancer Lett.* **1990**, *52*, 145–152.

596. Armitage, P.; Doll, R. The age distribution of cancer and a multistage theory of carcinogenesis. *Br. J. Cancer* **1954**, *8*, 1–12.

597. Fisher, J. C. Multiple mutation theory of carcinogenesis. *Nature* **1958**, *181*, 651–652.

598. Moolgavkar, S. H.; Knudson, A. G. Jr. Mutation and cancer: A model for human carcinogenesis. *J. Natl. Cancer Inst.* **1981**, *66*, 1037–1052.

599. Jackson, T.; Allard, M. F.; Sreenan, C. M.; Doss, L. K.; Bishop, S. P.; Swain, J. L. The c-*myc* protooncogene regulates cardiac development in transgenic mice. *Mol. Cell. Biol.* **1990**, *10*, 3709–3716.

600. Chang, J. D.; Billings, P. C.; Kennedy, A. R. C-*myc* expression is reduced in antipain-treated proliferating C3H 10T1/2 cells. *Biochem. Biophys. Res. Commun.* **1985**, *133*, 830–835.

601. Chang, J. D.; Kennedy, A. R. Cell cycle progression of C3H 10T1/2 and 3T3 cells in the absence of an increase in c-*myc* RNA levels. *Carcinogenesis* **1988**, *9*, 17–20.

602. Pfeifer-Ohlsson, S.; Goustin, A. S.; Rydnert, J.; et al. Spatial and temporal pattern of cellular *myc* oncogene expression in developing human placenta: implications for embryonic cell proliferation. *Cell* **1984**, *38*, 585–596.

603. Pfeifer-Ohlsson, S.; Rydnert, J.; Goustin, A. S.; Larsson, E.; Betsholtz, C.; Ohlsson, R. Cell-typespecific pattern of *myc* protooncogene expression in developing human embryos. *Proc. Natl. Acad. Sci. USA* **1985**, *82*, 5050–5054.

604. Downs, K. M.; Martin, G. R.; Bishop, J. M. Contrasting patterns of *myc* and N-*myc* expression during gastrulation of the mouse embryo. *Genes Dev.* **1989**, *3*, 860–869.

605. Schmid, P.; Schulz, W. A.; Hameister, H. Dynamic expression pattern of the *myc* protooncogene in midgestation mouse embryos. *Science* **1989**, *243*, 226–229.

606. Himing, U.; Schmid, P.; Schulz, W. A.; Rettenberger, G.; Hameister, H. A comparative analysis of N-*myc* and c-*myc* expression and cellular proliferation in mouse organogenesis. *Mechn. Dev.* **1991**, *33*, 119–126.

607. Birnie, G. D.; Warnock, A. M.; Burns, J. H.; Clark, P. Expression of the *myc* gene locus in populations of leukocytes from leukaemia patients and normal individuals. *Leukemia Res.* **1986**, *10*, 515–526.

608. Rothberg, P. G.; Erisman, M. D.; Diehl, R. E.; Rovigatti, U. G.; Astrin, S. M. Structure and expression of the oncogene c-*myc* in fresh tumor material from patients with hematopoietic malignancies. *Mol. Cell. Biol.* **1984**, *4*, 1096–1103.

609. Ferrari, S.; Torelli, U.; Selleri, L.; et al. Study of the levels of expression of two oncogenes, c-*myc* and c-*myb*, in acute and chronic leukemias of both lymphoid and myeloid lineage. Leukemia Res. **1985**, *9*, 833–842.

610. Momand, J.; Zambeffi, G. P.; Olson, D. C.; George, D.; Levine, A. J. The mdm-2 oncogene product forms a complex with the p53 protein and inhibits p53 mediated transactivation. *Cell* **1992**, *69*, 1237–1245.

611. Mukherjee, B.; Morgenbesser. S. D.; DePinho, R. A. Myc family oncoproteins function through a common pathway to transform normal cells in culture: cross-interference by Max and trans-acting dominant mutants. *Genes Dev.* **1992**, *6*, 1480–1492.
612. van Lohuizen, M.; Verbeek, S.; Krimpenfort, P.; et al. Predisposition to lymphomagenesis in pim-1 transgenic mice: cooperation with c-*myc* and N-*myc* in murine leukemia virus-induced tumors. *Cell* **1989**, *56*, 673–682.

CYTOGENETIC AND MOLECULAR STUDIES OF MALE GERM-CELL TUMORS

Eduardo Rodriguez, Chandrika Sreekantaiah, and

R. S. K. Chaganti

Advances in Genome Biology
Volume 3B, pages 415–428.
Copyright © 1995 by JAI Press Inc.
All rights of reproduction in any form reserved.
ISBN: 1-55938-835-8

I. INTRODUCTION

Adult male germ-cell tumors (GCTs) are a heterogeneous group of neoplasms that arise in premeiotic or early meiotic germ cells. They originate in gonadal or extragonadal sites (mediastinum, retroperitoneum, pineal) and histologically comprise two major groups: seminomas and nonseminomas. Seminomas are composed of neoplastic germ cells that mimic gametogenesis and act as immature spermatogenic cells. Nonseminomas present features of embryonic neoplastic germ cells and mimic histogenesis of the early embryo. Among nonseminomas, embryonal carcinoma is a pluripotential tumor which may progress along extraembryonic or trophoblastic lineages resulting in yolk sac tumor or choriocarcinoma, or along embryonic lineages resulting in teratoma. Individual tumors may present multiple histologies; mixed GCTs contain nonseminomatous components and combined GCTs present with seminomatous and nonseminomatous elements. Teratomatous lesions occasionally undergo malignant differentiation into tumors with other histologies such as sarcoma, carcinoma, neuronal or neuroectodermal tumor, or myeloid leukemia. The GCTs comprise a unique system for the study of the genetic basis of malignant transformation, differentiation, and drug sensitivity.

The GCTs are characterized by a highly specific chromosomal abnormality, i(12p), which is present in 76% of the tumors. Other nonrandom cytogenetic abnormalities are additional and are seen in lower frequencies. A subset of GCTs do not exhibit the i(12p) marker; in these other abnormalities leading to multiple copies of 12p are usually present. A combination of cytogenetic analyses with molecular techniques such as fluorescent *in-situ* hybridization and Southern blotting have resulted in a better resolution of some of the chromosomal changes and led to the identification of sites of candidate tumor suppressor genes in these tumors. This chapter provides a review of the current state of knowledge of cytogenetic and molecular cytogenetic studies of male GCTs and a correlation of these findings with the histological and clinical features of the tumors.

II. KARYOTYPIC PROFILE OF GERM-CELL TUMORS

Primary Chromosome Change.   Atkin and Baker first (1982) described the association of testicular GCTs with a specific chromosomal abnormality, i(12p). Since then 225 tumors have been cytogenetically analyzed and an i(12p) has been consistently reported in 76% of the tumors (Table 1). Despite the heterogeneity of the neoplasm, i(12p) has been characteristically observed in male GCTs of all histologies including seminomas, teratomas, embryonal carcinomas, choriocarcinomas, mixed and combined tumors from gonadal as well as extragonadal sites (Table 1). The data for Table 1 represent a combination of data from single case reports and cytogenetic series reported from different parts of the world; therefore, the incidence frequencies must be considered an approximation. The information,

Table 1. Incidence of i(12p) in Histologic Subsets of GCTs Detected by Cytogenetic Analysis[1,2,3]

Histologic Subset	Gonadal with i(12p)	Gonadal without i(12p)	Extragonadal with i(12p)	Extragonadal without i(12p)	Gonadal & Extragonadal with i(12p)	Gonadal & Extragonadal without i(12p)	Total
Seminoma	24(67)	12(33)	2(100)	0	26(68)	12(32)	38
Teratoma	61(88)	8(12)	4(100)	0	65(89)	8(11)	73
Embryonal carcinoma	20(77)	6(23)	0	1(100)	20(74)	7(26)	27
Choriocarcinoma	1(25)	3(75)	1(100)	0	2(40)	3(60)	5
Yolk sac tumor	3(100)	0	4(57)	3(43)	7(70)	3(30)	10
Non-seminoma with malignant transformation	6(100)	0	4(67)	2(33)	10(83)	2(17)	12
Mixed GCT	29(78)	8(22)	1(50)	1(50)	29(76)	9(24)	38
Combined GCT	5(42)	7(58)	1(100)	0	6(46)	7(54)	13
Undifferentiated Carcinoma	0	0	5(83)	1(17)	5(83)	1(17)	6
Pineal	0	0	0	3(100)	0	3(100)	3
Total	149(73)	44(27)	21(66)	11(34)	170(76)	52(24)	225

Notes: [1]Including primary and metastatic lesions.

[2]Based on the data reported in the following publications and unpublished results on 51 tumor biopsies from our laboratory: Albrecht et al., 1993 (1 tumor biopsy); Atkin and Baker, 1983 (4 tumor biopsies); Atkin et al., 1993 (3 tumor biopsies); Castedo et al., 1988a,b, 1989a,b,c (42 tumor biopsies); Dal Cin et al., 1989 (1 tumor biopsy); Delozier-Blanchet, 1985 (4 tumor biopsies); Gibas et al., 1984, 1986 (7 tumor biopsies); Haddad et al., 1988 (2 tumor biopsies); Hecht et al., 1984 (1 tumor biopsy); Murty et al., 1990 (7 cell lines); Oosterhuis et al., 1986, 1989, 1991 (4 tumor biopsies); Parrington et al., 1987 (4 cell lines); Rodriguez et al., 1991, 1992a, 1993 (58 tumor biopsies); Samaniego et al., 1990 (24 tumor biopsies); Shen et al., 1990 (1 tumor biopsy); Suijkerbuijk et al., 1993 (9 tumor biopsies, 1 cell line); Walt et al., 1986 (1 tumor biopsy).

[3]Numbers in parentheses indicate percentages.

nevertheless, is comparable to the findings on large series of tumors from single institutions in the published literature (de Jong et al., 1990; Rodriguez et al., 1992a). Although not much is known about the exact nature and role of i(12p) in the development of GCTs, the frequent occurrence of this marker, clearly reflects the significance of i(12p) in the development of this neoplasm.

Additional Nonrandom Chromosomal Changes. The modal chromosome number in GCTs is usually in the hyperdiploid to hypertetraploid range and numerous additional structural rearrangements are common. Based on the cytogenetic reports of the 225 tumors with clonal karyotypic abnormalities, the chromosomes most frequently rearranged, in addition to i(12p), were 1, 3, 6, 7, 9, 11, 12, and 17. A clustering of breakpoints to certain chromosomal sites and an association with histological or clinical features of tumors has been reported earlier (Rodriguez et al., 1992a). Thus, breaks at 1p32-36 and 7q11.2 were noted more frequently in teratomas, while 1p22 breaks were more frequent in yolk sac tumors. Breaks at 10p13, 7q11.2, and 12p11-q13 were more frequently encountered in metastatic lesions compared to primary lesions and breaks involving 1p36, 6q21, 7q13, and 12q11-13 were more frequent in posttreatment specimens compared to pretreatment specimens.

Cytogenetic Evidence of Gene Amplification. Homogeneously staining regions (HSRs) and double minute chromosomes (dmins) represent chromosomal manifestations of gene amplification, a phenomenon commonly associated with resistance to treatment of cells *in vitro* or tumors *in vivo* with antineoplastic drugs (Schwab and Amler, 1990). In the case of GCTs, presence of HSRs and molecular evidence for DNA amplification in HSR-bearing tumors was first presented by Samaniego et al. (1990). HSRs or dmins have been identified in 13 cases of GCTs, all but one of which were metastatic lesions (Rodriguez et al., 1992a; Albrecht et al., 1993). In one of these tumors, although DNA amplification was confirmed by DNA in-gel renaturation, amplification of any gene previously shown to be amplified in multiple tumor systems was excluded by hybridization of tumor DNA with a large panel of probes (Samaniego et al., 1990). Thus, it appears that a novel gene(s) is uniquely amplified in association with malignant progression of GCTs. The isolation and characterization of such gene(s) should be of significance not only to GCTs but to other tumors as well.

III. MOLECULAR ANALYSIS OF GERM-CELL TUMORS

Molecular Methods for Detection of i(12p). By conventional cytogenetic analysis, clonally abnormal karyotypes can be detected only in 50 to 70% of cases (Samaniego et al., 1990; Rodriguez et al., 1992a). Thus, in a substantial proportion of the cases, the status of chromosome 12 remains undetermined. One method to determine the copy number of 12p and/or 12q was by Southern blot analysis of tumor DNA hybridized with probes for loci on the respective arms. In this assay, the ratio of the signal between the target gene (a gene on 12p) and a reference gene (a gene on any chromosome arm other than 12p) in tumor DNA is compared to the identical ratio obtained from analysis of normal DNA (control), such as placental DNA. Increase in copy number of a 12p locus in tumor DNA relative to control DNA would indicate presence of multiple copies of 12p and hence, probably, i(12p). Analysis of such DNA ratios in cytogenetically characterized tumors and tumor cell lines showed an excellent correlation between DNA and cytogenetic assays for i(12p) (Dmitrovsky et al., 1990; Samaniego et al., 1990).

Identification of Candidate Tumor Suppressor Genes from Deletions in 12q. The detection of specific chromosomal deletions has classically enabled the identification of tumor suppressor genes: e.g., *RB, WT1, APC,* and *TP53* genes (Marshall, 1991). Cytogenetic studies in GCTs have identified deletions in 12q in up to 15% of GCTs, with or without simultaneous i(12p) (Castedo et al., 1989b; Murty et al., 1990; Samaniego et al., 1990; Rodriguez et al., 1992a, 1993). In order to determine whether the cytogenetic deletions in 12q truly represent molecular deletions, a recent study undertook analysis of loss of heterozygosity (LOH) comparing germline and tumor genotypes at eight polymorphic loci mapped to 12q. These results showed high frequency of LOH (> 40%) at two sites: 12q13 and 12q22, suggesting the presence of two candidate tumor suppressor genes at these regions (Murty et al., 1992). The presence of candidate tumor suppressor genes on 12q and the high frequency of i(12p) in these tumors highlight the central role of chromosome 12 in the development of male GCTs.

IV. MOLECULAR CYTOGENETICS OF GERM-CELL TUMORS

A simpler method than Southern blot analysis to determine abnormalities involving chromosome 12, mainly 12p, in GCTs utilizes the fluorescence *in situ* hybridization (FISH) technique. In this procedure, a biotinylated probe is hybridized directly on to metaphase chromosomes and/or nuclear preparations and visualized by fluorescence microscopy following incubation with fluoresceinated avidin and biotinylated goat antiavidin antibody and staining with an appropriate fluorochrome (Pinkel et al., 1986). Analysis by FISH with a chromosome 12 centromere-specific satellite DNA probe showed that the centromeres of the i(12p) chromosomes could be reliably distinguished from those of the normal chromosomes 12 by virtue of their larger or smaller sizes in tumor cells at metaphase as well as at interphase (Mukherjee et al., 1991; Rodriguez et al., 1992b). Parallel cytogenetic and FISH studies of a panel of tumors showed an excellent correlation between the two methods, thereby providing a rapid method of detection of this important marker in tumors and eliminating the limitations of conventional cytogenetic analysis (Rodriguez et al., 1992b).

A further enhancement of the FISH technique is chromosome painting in which the probe comprises pooled DNA fragments derived from an entire chromosome or chromosome arm. FISH analysis using such pooled probes recognizes ("paints") the corresponding chromosome or chromosome region (Figure 1). Using the painting technique, a high proportion of GCTs without i(12p) were shown to be characterized by an increased copy number of 12p incorporated into marker chromosomes proving that excess copy number of 12p is more frequent than was indicated by conventional cytogenetic or FISH analysis using the centromeric probe (Rodriguez et al., 1993; Suijkerbuijk et al., 1993) (Figure 1). Such studies also enhance the diagnostic value of this marker.

A

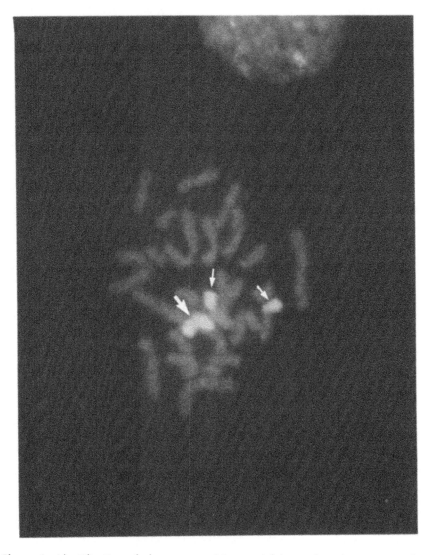

Figure 1. Identification of chromosome 12 material in marker chromosomes by chromosome painting using a mixed probe comprising DNA sequences isolated from chromosome 12. A is a partial metaphase from tumor 225A hybridized to a 12p painting probe showing i(12p) signal (*big arrow*) and normal chromosome 12 signals (*small arrow*). B is a partial metaphase from tumor 240A hybridized to a whole chromosome 12 painting probe showing signal in two markers (big arrows) and two normal 12 (small arrows). C is a partial metaphase from tumor 268A hybridized to 12p painting probe showing signal in two markers (arrows).

B

Figure 1. (continued)

C

Figure 1. (continued)

V. APPLICATION OF CYTOGENETIC FINDINGS IN THE DIAGNOSIS AND PROGNOSIS OF GERM-CELL TUMORS

Because of its high incidence in GCTs, i(12p) has been shown to be a valuable diagnostic marker in several types of histologically ambiguous situations. Thus, a syndrome of acute myeloid leukemia (AML) associated with a very poor prognosis resulting form malignant hematopoietic transformation of i(12p)-bearing teratomatous cells has been identified (Chaganti et al., 1989; Ladanyi et al., 1990; Nichols et al., 1990). Therefore, AML arising in patients with a prior or concomitant history of GCT must be evaluated for germ-cell origin of the leukemia. A primary diagnosis of GCT can also be established by this marker in patients with a diagnosis of poorly differentiated carcinomas or adenocarcinomas of unknown primary sites. Such patients generally do poorly with systemic chemotherapy (Didolkar et al., 1977; Woods et al., 1980). In a minority of these patients who also presented with the clinical features of the unrecognized extragonadal germ-cell cancer syndrome (UEGCCS) (age <50 years, tumors involving midline structures, lung parenchyma, or lymph nodes), long-term survival with cisplatin-based chemotherapy was achieved (Richardson et al., 1981; Greco et al., 1982, 1986). In a recent study of nine patients with UEGCCS, four were shown to have abnormalities of chromosome 12 consistent with a diagnosis of GCT (Motzer et al., 1991). Therefore, patients with undifferentiated carcinomas of unknown primary, especially those involving the midline structures, must be subjected to cytogenetic analysis to rule out a diagnosis of GCT. Additionally, the use of powerful new techniques, such as those described above, now make it possible to determine precisely the number of 12p copies and verify the initial observations of Bosl et at. (1989) which suggested that the copy number of this chromosomal arm may predict survival.

VI. CYTOGENETIC BASIS OF MALIGNANT TRANSFORMATION IN TERATOMATOUS LESIONS

A number of teratomatous lesions undergo malignant transformation presenting specific differentiation patterns of nongerm cell lineages such as sarcoma, adenocarcinoma, neuroepilthelioma, and myeloid leukemia; these cases have been shown to be accompanied by the appearance of specific chromosomal changes which have been previously characterized in *de novo* tumors of the same histology (Ulbright et al., 1984; Chaganti et al., 1989; Ladanyi et al., 1990; Rodriguez et al., 1991, 1992a). The origin of this transformation, in particular its clonal relationship to the teratomatous lesion, has been determined in a number of cases. One example is a patient in whom myeloid leukemia presented a mediastinal teratoma and yolk sac tumor in less than 2 years from the diagnosis. Both lesions were shown to be clonally related by detection of i(12p); the leukemia in addition acquired a deletion in 5q which is characteristic of *de novo* myeloid leukemias (Chaganti et al., 1989).

Similarly, malignant transformation of teratomatous lesions to embryonal rhab-domyosarcoma and neuroepilthelioma were shown to be associated with 2q37 and 11q24 rearrangements, respectively—aberrations which are characteristic of *de novo* tumors with the same histologies (Rodriguez et al., 1991; 1992a). The elucidation of the cytogenetic basis of these transformations not only comprises additional support to the view that specific chromosome changes are etiologically relevant in tumorigenesis, but also provides new opportunities for isolation of genes at these sites.

VII. CHROMOSOME ABNORMALITIES AND PATHOGENESIS OF GERM-CELL TUMORS

Testicular seminomas as well as nonseminomas are suggested to develop from carcinoma *in situ* (CIS) (Skakkabaek et al., 1987), while primary mediastinal GCTs are considered to arise from primordial germ cells held behind in the midline during their embryonal migration from yolk sac to the gonadal ridge (Ashley, 1979; Gonzales-Crussi, 1982). However, due to the fact that mediastinal GCTs exhibit identical nonrandom chromosomal changes as gonadal GCTs, it has been proposed that all GCTs have a gonadal origin (Chaganti et al., 1993); occasional migration of GCT precursors early in their development to extragonadal sites would become established as primary mediastinal GCTs. Gonadal seminomas exhibit chromo-some numbers in the hypertriploid to tetraploid ranges, and nonseminomas present chromosome numbers in the hypotriploid ranges (Atkin, 1973; de Jong et al., 1990). In addition, nonseminomas have been shown to be predominantly XXY in sex chromosome constitution (Atkin et al., 1991). Based on these data, polyploizization has been suggested to be the initial cytogenetic change in the pathogenesis of this tumor (Vos et al., 1990; Atkin et al., 1991). de Jong et al (1990) proposed a model of pathogenesis which called for an initial polyploidization of the precursor cell resulting in CIS whose further evolution to seminomas and nonseminoma would be based on generation of the i(12p) marker and progressive and selective loss of chromosomes by nondisjunction. According to this model, the i(12p) formation would be a secondary event possibly related to progression. However, since 12p amplification has been shown to be present in i(12p)-negative tumors as well (Rodriguez et al., 1993; Suijkerbuijk et al., 1993), this suggests that this chromo-somal change must be involved in the initiation process. A model for the genesis of GCTs has been proposed (Chaganti et al., 1993) in which a pachytene–diplotene spermatocyte is the precursor cell for all GCTs. If in an occasional pachytene–diplotene cell the repair mechanism, necessary to complete crossingover, fail due to failure of induction of a key protein, the cell would be unable to progress further in meiosis and degenerates. A sporadic meiocyte with a defective repair mechanism may be rescued from death by initiation of a new program of mitotic division. Therefore, the precursor GCT cell could be a 4C cell with defective repair aberrant-exchange events affecting the pericentromeric regions of 12p leading to i(12p), or tandem duplication of 12p segments providing the necessary extra copy number (and the increased product) of the key gene which initiates the new cycling event and rescues the cell from death. Initiation of one round of DNA replication in the 4C cell would lead to endoreduplication and a tetraploid cell with an i(12p) or increased 12p copies. Because of the initial defective repair, the cell may be genetically unstable and liable to undergo additional changes such as nondisjunction, deletion, and mutation.

The potential significance of chromosome 12 alterations in GCT development has been alluded to above. The data comprise is: (1) i(12p) is common to all histologic subsets and anatomic presentations of GCTs, unrelated to ploidy levels of individual tumors; (2) most tumors with i(12p) [including those with 2 normal copies of chromosome 12 and one i(12p)] retain both paternal and maternal copies of chromosome 12, suggesting that i(12p) is generated by either exchanges in homologous nonsister chromatids as suggested by Mukherjee et al. (1991) or by unequal "misdivision" following a nondisjunctional gain of one of the chromosomes 12 in the precursor cell; and (3) deletion analysis provided evidence for two candidate tumor suppressor genes on 12q that are potentially unique to this tumor system (Murty et al., 1992). Therefore, chromosome 12 alterations are of critical significance in the development of these tumors.

VIII. SUMMARY

Although the biological significance of GCTs to the study of malignancy and differentiation has been well recognized for some time, detailed genetic analysis based on fresh tumor biopsies has not been initiated until recently. The first stage of such studies, as with other, more extensively investigated tumor systems, has been cytogenetic analysis. To date, cytogenetic data on 225 tumors are available which, although small in number, have already yielded valuable insights into the biology of these tumors and a clinically useful marker. Thus, initial correlations between chromosome change and histology have been recorded, gene amplification associated with malignant progression has been identified, cytogenetic basis of malignant differentiation in teratomatous lesions has been clarified, and sites of candidate tumor suppressor genes unique to this system have been identified. The usefulness of i(12p) as a diagnostic marker, especially in tumors of uncertain histologies has been established. Because of the clinical usefulness of this marker, molecularly based methods for its detection, without the need for formal cytogenetic analysis, have been developed. Cytogenetic analysis of larger prospectively ascertained series than have been performed so far and analysis of large numbers of tumors utilizing molecular techniques can be expected to yield significant insights into the biology and clinical behavior of these tumors.

REFERENCES

1. Albrecht, S.; Armstrong, D. L.; Mahoney, D. H.; et al. Cytogenetic demonstration of gene amplification in a primary intracranial germ cell tumor. Genes Chrom. *Cancer* **1993**, *6*: 61–63.
2. Ashley, D. J. B. Origin of teratomas. *Cancer* **1973**, *32*, 390–394.
3. Atkin, N. B.; Baker, M. C. Specific chromosome change, i(12p), in testicular tumors? *Lancet* **1982**, *2*, 1349.
4. Atkin, N. B.; Baker, M.C. i(12p): Specific chromosomal marker in seminoma and malignant teratoma of the testis? *Cancer Genet. Cytogenet.* **1983**, *10*, 199–204.
5. Atkin, N. B.; Baker, M. C. X-chromatin, sex chromosomes, and ploidy in 37 germ cell tumors of the testis. *Cancer Genet. Cytogenet* **1992**, *59*, 54–56.
6. Atkin, N. B.; Fox, M. F.; Baker, M. C.; et al. Chromosome 12-containing markers, including two dicentrics, in three i(12p)-negative testicular germ cell tumors. *Genes Chrom. Cancer* **1993**, *6*, 218–221.
7. Bosl, J. G.; Dmitrovsky, E.; Reuter, V. E.; et al. Isochromosome of chromosome 12p: a clinically useful chromosomal marker for male germ cell tumors. *J. Natl. Cancer Inst.* **1989**, *18*, 1874–1878.
8. Castedo, S. M. M. J.; de Jong, B.; Oosterhuis, J. W.; et al. Cytogenetic study of a combined germ cell tumor of the testis. *Cancer Genet. Cytogenet.* **1988a**, *35*, 159–165,
9. Castedo, S. M. M. J.; de Jong, B.; Oosterhuis, J. W.; et al. i(12p)-Negative testicular germ cell tumors. A different group? *Cancer Genet. Cytogenet.* **1988b**, *35*, 171–178.
10. Castedo, S. M. M. J.; de Jong, B.; Oosterhuis, J. W.; et al. Cytogenetic analysis of ten human seminomas. *Cancer Res.* **1989a**, *49*, 439–443.
11. Castedo, S. M. M. J.; de Jong, B.; Oosterhuis, J. W.; et al. Chromosomal changes in mature residual teratomas following polychemotherapy. *Cancer Res.* **1989b**, *49*, 672–676.
12. Castedo, S. M. M. J.; de Jong, B.; Oosterhuis, J. W.; et al. Chromosomal changes in human primary testicular nonseminomatous germ cell tumors. *Cancer Res.* **1989c**, *49*, 5696–5701.
13. Chaganti, R. S. K.; Ladanyi, M.; Samaniego, F.; et al. Malignant hematopoietic differentiation of a germ cell tumor. *Genes Chrom. Cancer* **1989**, *1*, 83–87.
14. Chaganti, R. S. K.; Rodriguez, E.; Mathew, S. Mediastinal germ cell tumors in males: the case for a testicular origin. Lancet, in press.
15. Chaganti, R. S. K.; Murty, V. V. V. S.; Houldsworth, J.; et al. Molecular genetics in germ cell tumor. In: *Germ Cell Tumors III* (Jones G., Ed.). W.B. Sanders, New York, in press.
16. Dal Cin, P.; Drochmans, A.; Moerman, P.; et al. Isochromosome 12p in mediastinal germ cell tumor. *Cancer Genet. Cytogenet.* **1989**, *42*, 243–251.
17. Darlington, C. D. The origin of isochromosomes. *J. Genet.* **1940**, *39*, 351–361.
18. de Jong, B.; Oosterhuis, J. W.; Castedo, S. M. M. J.; et al. Pathogenesis of adult testicular germ cell tumors. A cytogenetic model. *Cancer Genet. Cytogenet.* **1990**, *48*, 143–167.
19. Delozier-Blanchet, C. D.; Engel, E.; Walt, H. Isochromosome 12p in malignant testicular tumors. *Cancer Genet. Cytogenet.* **1985**, *15*, 375–376.
20. Didolkar, M. S.; Fanous, N.; Elias, E. G.; et al. Metastatic carcinomas from occult primary tumors. A study of 254 patients. *Ann. Surg.* **1977**, *186*, 625–630.
21. Dmitrovsky, E.; Murty, V. V. V. S.; Moy, D.; et al. Isochromosome 12p in non-seminoma cell lines: Karyologic amplification of c-ki-*ras* 2 without point-mutational activation. *Oncogene* **1990**, *5*, 543–548.
22. Gibas, Z.; Prout, G. R.; Pontas, J. E.; et al. Chromosome changes in germ cell tumors of the testis. *Cancer Genet. Cytogenet.* **1986**, *19*, 245–252.
23. Gonzales-Crussi, F. Gonadal teratoma. *Atlas of Tumor Pathology*. Fascicle 18. Armed Forces Institute of Pathology, Washington, DC, 1982.
24. Greco, F. A.; Oldham, R. K.; Fer, M. F. The extragonadal germ cell cancer syndrome. *Semin. Oncol.* **1982**, *9*, 448–455.

25. Greco, F. A.; Vaughn, W. K.; Hainsworth, J. D. Advanced poorly differentiated carcinoma of unknown primary site: Recognition of a treatable syndrome. *Ann. Intern. Med.* **1986**, *104*, 547–553.

26. Greig, G. M.; Parikh, S.; George, J.; et al. Molecular cytogenetics of α-satellite DNA from chromosome 12: fluorescence *in situ* hybridization and description of DNA and assay length polymorhisms. *Cytogenet. Cell Genet.* **1991**, *56*, 144–148.

27. Haddad, F. S.; Sorini, P. M.; Somsin, A. A.; et al. Familial double testicular tumors: Identical chromosome changes in seminoma and embryonal carcinoma of the same testis. *J. Urol.* **1988**, *139*, 748–751.

28. Hecht, F.; Grix, A.; Hecht, B. K.; et al. Direct prenatal chromosome diagnosis of malignancy. *Cancer Genet. Cytogenet.* **1984**, *11*, 107–111.

29. Ladanyi, M.; Samaniego, F.; Reuter, V.; et al. Cytogenetic and immunohistochemical evidence for the germ cell origin of a subset of acute leukemias associated with mediastinal germ cell tumors. *J. Natl. Cancer Inst.* **1990**, *82*, 221–227.

30. Marshall, C. J. Tumor suppressor genes. *Cell* **1991**, *64*, 313–326.

31. Motzer, R. J.; Rodriguez, E.; Reuter, V. E.; et al. Genetic analysis as an aid in diagnosis for patients with midline carcinomas of uncertain histologies. *J. Natl. Cancer Inst.* **1991**, *83*, 341–346.

32. Mukherjee, A. B.; Murty, V. V. V. S.; Rodriguez, E.; et al. Detection and analysis of origin of i(12p), a diagnostic marker of human male germ cell tumors, by fluorescence *in situ* hybridization. *Genes Chrom. Cancer* **1991**, *3*, 300–307.

33. Murty, V. V. V. S.; Dmitrovsky, E.; Bosl, G. J.; et al. Nonrandom chromosome abnormalities in testicular and ovarian germ cell tumor cell lines. *Cancer Genet. Cytogenet.* **1990**, *50*, 67–73.

34. Murty, V. V. V. S.; Houldsworth, J.; Baldwin, S.; et al. Allelic deletions in the long arm of chromosome 12 identify sites of candidate tumor suppressor genes in male germ cell tumors. *Proc. Natl. Acad. Sci. USA* **1992**, *89*, 11006–11010.

35. Oosterhuis, J. W.; de Jong, B.; Cornelisse, C. J.; et al. Karyotyping and DNA flow cytometry of mature residual teratoma after intensive chemotherapy of disseminated nonseminomatous germ cell tumor of the testis: a report of two cases. *Cancer Genet. Cytogenet.* **1986**, *22*, 149–157.

36. Oosterhuis, J. W.; Castedo, S. M. M. J.; de Jong, B.; et al. A malignant mixed gonadal stromal tumor of the testis with heterologous components and i(12p) in one of its metastases. *Cancer Genet. Cytogenet.* **1989**, *41*, 105–114.

37. Oosterhuis, J. W.; van der Berg, E.; de Jong, B.; et al. Mediastinal germ cell tumor with secondary nongerm cell malignancy, and extensive hematopoietic activity. *Cancer Genet. Cytogenet.* **1991**, *54*, 183–195.

38. Parrington, J. M.; West, L. F.; Povey, S. Loss of heterozygosity in hypotriploid cell cultures from testicular tumors. *Hum. Genet.* **1987**, *77*, 269–276.

39. Pinkel, D.; Straume, T.; Gray, J. W. Cytogenetic analysis using quantitative, high sensitivity fluorescence hybridization. *Proc. Natl. Acad. Sci. USA* **1986**, *83*, 2934–2938.

40. Pugh, R. C. B. Combined tumors. In: *Pathology of the Testis* (Pugh, R.C.B., Ed.) pp. 245–258. Blackwell, Oxford, 1976.

41. Richardson, R. L.; Schoumacher, R. A.; Fer, M. F.; et al. The unrecognized extragonadal germ cell cancer syndrome. *Ann. Intern. Med.* **1981**, *94*, 181–186.

42. Rodriguez, E.; Reuter, V. E.; Mies, C.; et al. Abnormalities of 2q: a common genetic link between rhabdomyosarcoma and hepatoblastoma? *Genes Chrom. Cancer* **1991**, *3*, 122–127.

43. Rodriguez, E.; Mathew, S.; Reuter, V.; et al. Cytogenetic analysis of 124 prospectively ascertained male germ cell tumors. *Cancer Res.* **1992a**, *52*, 2285–2291.

44. Rodriguez, E.; Mathew, S.; Mukherjee, A. B.; et al. Analysis of chromosome 12 aneuploidy in interphase cells from human male germ cell tumors by fluorescence in situ hybridization. *Genes Chrom. Cancer* **1992b**, *5*, 21–29.

45. Rodriguez, E.; Houldsworth, J.; Reuter, V. E.; et al. Molecular cytogenetic analysis of i(12p)-negative human male germ cell tumors. *Genes Chromosomes Cancer* **1993**, *8*, 230–236.

46. Samaniego, F.; Rodriguez, E.; Houldsworth, J.; et al. Cytogenetic and molecular analysis of human male germ cell tumors: chromosome 12 abnormalities and gene amplification. *Genes Chrom. Cancer* **1990**, *1*, 289–300.

47. Shen, V.; Chaparro, M.; Choi, B. H.; et al. Absence of isochromosome 12p in a pineal region malignant germ cell tumor. *Cancer Genet. Cytogenet.* **1990**, *50*, 153–160.

48. Sledge, G. W., Jr.; Glant, M.; Jansen, J.; et al. Establishment in long term culture of megakaryo-cytic leukemia cells (EST-IU) from the marrow of a patient with leukemia and a mediastinal germ cell neoplasm. *Cancer Res.* **1986**, *46*, 2155–2159.

49. Ulbright, T. M.; Loehrer, P. J.; Roth, L. M.; et al. The development of non-germ cell malignancies within germ cell tumors. *Cancer* **1984**, *54*, 1824–1833.

50. Ulbright, T. M.; Roth, L. M. Recent developments in the pathology of germ cell tumors. *Semin. Diagn. Pathol.* **1987**, *4*, 304–319.

51. Vos, A. M.; Oosterhuis, J. W.; de Jong, B.; et al. Cytogenetics of carcinoma in situ of testis. *Cancer Genet. Cytogenet.* **1990**, *46*, 75–81.

52. Walt, H.; Arenbrecht, S.; Delozier-Blanchet, C. D.; et al. A human testicular germ cell tumor with border line histology between seminoma and embryonal carcinoma secreted beta-human chorionic gonadotropin and alpha-fetoprotein only as a xenograft. *Cancer* **1986**, *58*, 139–146.

53. Woods, R. L.; Fox, R. M.; Tattersall, M. H. N.; et al. Metastatic adenocarcinomas of unknown primary site: a randomized study of two combination chemotherapy regimens. *N. Engl. J. Med.* **1980**, *303*, 87–89.

INDEX

Printed and bound by CPI Group (UK) Ltd, Croydon, CR0 4YY

03/10/2024

01040436-0014